MILITARY AND VETERAN ISSUES

NAVY SHIPBOARD LASERS

BACKGROUND, ADVANCES, AND CONSIDERATIONS

MILITARY AND VETERAN ISSUES

Additional books in this series can be found on Nova's website under the Series tab.

Additional e-books in this series can be found on Nova's website under the e-book tab.

MILITARY AND VETERAN ISSUES

NAVY SHIPBOARD LASERS

BACKGROUND, ADVANCES, AND CONSIDERATIONS

WILLIAM C. JENKINS
EDITOR

nova publishers
New York

Library of Congress Cataloging-in-Publication Data

ISBN: 978-1-63483-372-1

Published by Nova Science Publishers, Inc. † New York

CONTENTS

Preface		**vii**
Chapter 1	Navy Shipboard Lasers for Surface, Air, and Missile Defense: Background and Issues for Congress *Ronald O'Rourke*	**1**
Chapter 2	Propagation of High Energy Laser Beams Through Atmospheric Stagnation Zones *Joseph R. Penano, Phillip Sprangle and Bahman Hafizi*	**71**
Index		**87**

PREFACE

Department of Defense (DOD) development work on high-energy military lasers, which has been underway for decades, has reached the point where lasers capable of countering certain surface and air targets at ranges of about a mile could be made ready for installation on Navy surface ships over the next few years. More powerful shipboard lasers, which could become ready for installation in subsequent years, could provide Navy surface ships with an ability to counter a wider range of surface and air targets at ranges of up to about 10 miles. This book focuses on potential Navy shipboard lasers for countering surface, air, and ballistic missile threats.

In: Navy Shipboard Lasers
Editor: William C. Jenkins
ISBN: 978-1-63483-372-1

Chapter 1

NAVY SHIPBOARD LASERS FOR SURFACE, AIR, AND MISSILE DEFENSE: BACKGROUND AND ISSUES FOR CONGRESS*

Ronald O'Rourke

SUMMARY

Department of Defense (DOD) development work on high-energy military lasers, which has been underway for decades, has reached the point where lasers capable of countering certain surface and air targets at ranges of about a mile could be made ready for installation on Navy surface ships over the next few years. More powerful shipboard lasers, which could become ready for installation in subsequent years, could provide Navy surface ships with an ability to counter a wider range of surface and air targets at ranges of up to about 10 miles.

The Navy and DOD have conducted development work on three principal types of lasers for potential use on Navy surface ships—fiber solid state lasers (SSLs), slab SSLs, and free electron lasers (FELs). One fiber SSL prototype demonstrator developed by the Navy is the Laser Weapon System (LaWS). The Navy in August 2014 installed a LaWS system on the USS *Ponce*, a ship operating in the Persian Gulf as an

* This is an edited, reformatted and augmented version of a Congressional Research Service publication, No. R41526, dated June 1, 2015.

interim Afloat Forward Staging Base (AFSB[I]), to conduct continued evaluation of shipboard lasers in an operational setting. The Navy reportedly anticipates moving to a shipboard laser program of record in "the FY2018 time frame" and achieving an initial operational capability (IOC) with a shipboard laser in FY2020 or FY2021.

Although the Navy is developing laser technologies and prototypes of potential shipboard lasers, and has a generalized vision for shipboard lasers, the Navy currently does not yet have a program of record for procuring a production version of a shipboard laser. The possibility of equipping Navy surface ships with lasers in coming years raises a number of potential issues for Congress, including the following:

- how many types of lasers to continue developing, particularly given constraints on Navy funding, and the relative merits of types currently being developed; and
- the potential implications of shipboard lasers for the design and acquisition of Navy ships, including the Flight III DDG-51 destroyer that the Navy wants to begin procuring in FY2016.

INTRODUCTION

Issue for Congress

Department of Defense (DOD) development work on high-energy military lasers, which has been underway for decades, has reached the point where lasers capable of countering certain surface and air targets at ranges of about a mile could be made ready for installation on Navy surface ships over the next few years. More powerful shipboard lasers, which could become ready for installation in subsequent years, could provide Navy surface ships with an ability to counter a wider range of surface and air targets at ranges of up to about 10 miles.

The Navy in August 2014 installed a prototype solid state laser called the Laser Weapon System (LaWS) on the USS *Ponce*, a ship operating in the Persian Gulf as an interim Afloat Forward Staging Base (AFSB[I]), to conduct continued evaluation of shipboard lasers in an operational setting. The Navy reportedly anticipates moving to a shipboard laser program of record in "the FY2018 time frame" and achieving an initial operational capability (IOC) with a shipboard laser in FY2020 or FY2021.[1]

Compared to existing ship self-defense systems, such as missiles and guns, lasers could provide Navy surface ships with a more cost effective means of countering certain surface, air, and ballistic missile targets. Ships equipped with a combination of lasers and existing self-defense systems might be able to defend themselves more effectively against a range of such targets. Equipping Navy surface ships with lasers could lead to changes in naval tactics, ship design, and procurement plans for ship-based weapons, bringing about a technological shift for the Navy—a "game changer"—comparable to the advent of shipboard missiles in the 1950s.

The central issue for Congress is whether to approve or modify the Administration's proposed funding levels for development of potential shipboard lasers, and whether to provide the Navy or DOD with direction concerning development and procurement programs for shipboard lasers. Potential specific issues for Congress include the following:

- how many types of lasers to continue developing, particularly given constraints on Navy funding, and the relative merits of types currently being developed; and
- the potential implications of shipboard lasers for the design and acquisition of Navy ships, including the Flight III DDG-51 destroyer that the Navy wants to begin procuring in FY2016.

Decisions that Congress makes regarding potential shipboard lasers could significantly affect future Navy capabilities and funding requirements, the U.S. industrial base for military lasers, and the industrial base for existing shipboard self-defense systems.

Scope, Sources, and Terminology

This report focuses on potential Navy shipboard lasers for countering surface, air, and ballistic missile threats. It does not discuss the use of lasers on Navy aircraft or submarines, or the use of lasers by other military services. This report is based on unclassified, open-domain information from the Navy, industry, and research organizations such as RAND.

For purposes of this report, the term "short range" generally refers to ranges of one or two nautical miles, while references to longer ranges or extended ranges refer to ranges of up to about 10 nautical miles.[2] Lasers are one type of directed energy weapon (DEW); other DEWs include microwave weapons and millimeter wave weapons. (Another new weapon being developed by the Navy, the electromagnetic rail gun, is an electrically powered weapon, but strictly speaking is not a directed energy weapon, since it fires a projectile.)

Background

Shipboard Lasers in General

Potential Advantages and Limitations of Shipboard Lasers

Lasers are of interest to the Navy and other observers as potential shipboard weapons because they have certain potential advantages for countering some types of surface, air, and ballistic missile targets. Shipboard lasers also have potential limitations for countering such targets. Potential advantages and limitations are discussed below.

Advantages

Potential advantages of shipboard lasers for countering surface, air, and ballistic missile targets include the following:

- **Low marginal cost per shot.** Shipboard lasers could counter surface, air, and ballistic missile targets at a low marginal cost per shot. The shipboard fuel needed to generate the electricity for firing an electrically powered laser would cost less than a dollar per shot (some sources express the cost in pennies per shot).[3] In contrast, the Navy's short-range air-defense interceptor missiles cost hundreds of thousands (or more than a millions dollars) each, and its longer-range air- and missile-defense interceptor missiles cost several million dollars each. A laser can give a ship an alternative to using an expensive interceptor missile to achieve a "hard kill"[4] against a much less expensive target, such as an unsophisticated unmanned air vehicle (UAV). A low marginal cost per shot could permit the Navy to dramatically improve the cost exchange ratio—the cost of the attacker's weapon compared to the Navy's marginal cost per shot for countering that weapon. Cost exchange ratios currently often favor the attacker, sometimes very significantly. Converting unfavorable cost exchange ratios into favorable ones could be critical for the Navy's ability in coming years to mount an affordable defense against adversaries that choose to deploy large numbers of small boats,

UAVs, anti-ship cruise missiles (ASCMs), and anti-ship ballistic missiles (ASBMs) for possible use against U.S. Navy ships.

- **Deep magazine.** Navy surface ships can carry finite numbers of interceptor missiles in their missile launch tubes. Once a Navy surface ship's interceptors are fired, loading a new set of interceptors onto the ship would require the ship to temporarily withdraw from the battle. The Phalanx Close-In Weapon System (CIWS) that is installed on Navy surface ships—a radar-controlled Gatling gun that fires bursts of 20mm shells—similarly can engage a finite number of targets before it needs to be reloaded, which takes a certain amount of time. In contrast, an electrically powered laser can be fired again and again, as long as the ship has fuel to generate electricity (and sufficient cooling capacity to remove waste heat from the laser). A laser would give a ship a weapon with a deep (some observers say virtually unlimited) magazine capacity. Lasers could permit Navy surface ships to more effectively defend themselves against adversaries with more weapons and decoys than can be handled by the ships' onboard supplies of interceptor missiles and CIWS ammunition. A ship equipped with a laser, for example, could use the laser to counter an initial wave of decoys while conserving the ship's finite supply of interceptor missiles and CIWS ammunition for incoming weapons that are best countered by those systems. Future ships designed with a combination of lasers and missile-launch tubes could be smaller, and thus less expensive to procure, than future ships designed with no lasers and a larger number of missile-launch tubes.
- **Fast engagement times.** Light from a laser beam can reach a target almost instantly (eliminating the need to calculate an intercept course, as there is with interceptor missiles) and, by remaining focused on a particular spot on the target, cause disabling damage to the target within seconds. After disabling one target, a laser can be redirected in several seconds to another target. Fast engagement times can be particularly important in situations, such as near-shore operations, where missiles, rockets, artillery shells, and mortars could be fired at Navy ships from relatively close distances.
- **Ability to counter radically maneuvering air targets.** Lasers can follow and maintain their beam on radically maneuvering air targets (such as certain ASCMs) that might stress the maneuvering capabilities of Navy interceptor missiles.

- **Precision engagement and reduced risk of certain kinds of collateral damage in port areas.** Lasers are precision-engagement weapons—the light spot from a laser, which might be several inches in diameter, affects what it hits, while generally not affecting (at least not directly) separate nearby objects. Navy ships in overseas ports might be restricted in their ability to use the CIWS to defend themselves against mortars and rockets out of concern that CIWS shells that are fired upward but miss the target would eventually come back down, possibly causing collateral damage in the port area. In contrast, light from an upward-pointing laser that does not hit the target would continue flying upward in a straight line, which can reduce the chance of causing collateral damage to the port area.
- **Additional uses; graduated responses.** Lasers can perform functions other than destroying targets, including detecting and monitoring targets and producing nonlethal effects, including reversible jamming of electro-optic (EO) sensors.[5] Lasers offer the potential for graduated responses that range from warning targets to reversibly jamming their systems, to causing limited but not disabling damage (as a further warning), and then finally causing disabling damage.

Limitations

Potential limitations of shipboard lasers for countering surface, air, and ballistic missile targets include the following:

- **Line of sight.** Since laser light tends to fly through the atmosphere on an essentially straight path, shipboard lasers would be limited to line-of-sight engagements, and consequently could not counter over-the-horizon targets or targets that are obscured by intervening objects. This limits in particular potential engagement ranges against small boats, which can be obscured by higher waves, or low-flying targets. Even so, lasers can rapidly reacquire boats obscured by periodic swells, and more generally might be able to engage targets at longer ranges than certain existing shipboard gun systems. An airborne mirror, perhaps mounted on an aerostat,[6] could bounce light from a shipboard laser, so as to permit non-line-of-sight engagements; implementing such an arrangement would add cost and technical challenges, and the aerostat could be damaged by a misaimed shipboard laser or enemy attack.

- **Atmospheric absorption, scattering, and turbulence; not an all-weather solution.** Substances in the atmosphere—particularly water vapor, but also things such as sand, dust, salt particles, smoke, and other air pollution—absorb and scatter light from a shipboard laser, and atmospheric turbulence can defocus a laser beam. These effects can reduce the effective range of a laser. Absorption by water vapor is a particular consideration for shipboard lasers because marine environments feature substantial amounts of water vapor in the air.[7] There are certain wavelengths of light (i.e., "sweet spots" in the electromagnetic spectrum) where atmospheric absorption by water vapor is markedly reduced.[8] Lasers can be designed to emit light at or near those sweet spots, so as to maximize their potential effectiveness. Absorption generally grows with distance to target, making it in general less of a potential problem for short-range operations than for longer-range operations. Adaptive optics, which make rapid, fine adjustments to a laser beam on a continuous basis in response to observed turbulence, can counteract the effects of atmospheric turbulence. Even so, lasers might not work well, or at all, in rain or fog, preventing lasers from being an all-weather solution.
- **Thermal blooming.** A laser that continues firing in the same exact direction for a certain amount of time can heat up the air it is passing through, which in turn can defocus the laser beam, reducing its ability to disable the intended target. This effect, called thermal blooming, can make lasers less effective for countering targets that are coming straight at the ship, on a constant bearing (i.e., "down-thethroat" shots). Other ship self-defense systems, such as interceptor missiles or a CIWS, might be more suitable for countering such targets. Most tests of laser systems have been against crossing targets rather than "down-the-throat" shots. In general, thermal blooming becomes more of a concern as the power of the laser beam increases.
- **Saturation attacks.** Since a laser can attack only one target at a time, requires several seconds to disable it, and several more seconds to be redirected to the next target, a laser can disable only so many targets within a given period of time. This places an upper limit on the ability of an individual laser to deal with saturation attacks—attacks by multiple weapons that approach the ship simultaneously or within a few seconds of one another. This limitation can be mitigated by installing more than one laser on the ship, similar to how the Navy installs multiple CIWS systems on certain ships.[9]

- **Hardened targets and countermeasures.** Less-powerful lasers—that is, lasers with beam powers measured in kilowatts (kW) rather than megawatts (MW)[10]— can have less effectiveness against targets that incorporate shielding, ablative material, or highly reflective surfaces, or that rotate rapidly (so that the laser spot does not remain continuously on a single location on the target's surface) or tumble.[11] Small boats could employ smoke or other obscurants to reduce their susceptibility to laser attack. Measures such as these, however, can increase the cost and/or weight of a weapon, and obscurants could make it more difficult for small boat operators to see what is around them, reducing their ability to use their boats effectively.
- **Risk of collateral damage to aircraft and satellites.** Since light from an upward-pointing laser that does not hit the target would continue flying upward in a straight line, it could pose a risk of causing unwanted collateral damage to aircraft and satellites.[12]

In addition to the above points, a shipboard laser, like other shipboard systems, would take up space on a ship, use up some of the ship's weight-carrying capacity, create a load on the ship's electrical power and cooling systems, and possibly alter the ship's radar cross section. These considerations—referred to collectively as ship impact—can become significant when considering whether to backfit lasers onto existing ships, or whether to incorporate lasers into new ship designs.[13]

Potential Targets for Shipboard Lasers

Potential targets for shipboard lasers include the following:

- electro-optical (EO) sensors, including those on anti-ship missiles;
- small boats (including so-called "swarm boats")[14] and other watercraft (such as jet skis);
- rockets, artillery shells, mortars (sometimes collectively referred to as RAM);
- UAVs;
- manned aircraft;
- ASCMs; and
- ballistic missiles, including ASBMs.

Small boats, rockets, artillery shells, and mortars can be a particular concern for Navy surface ships during operations close to shore. Iran has acquired large numbers of swarm boats for potential use during a crisis or conflict against U.S. Navy ships seeking to enter or operate in the Persian Gulf. RAM weapons are widely proliferated to both state and non-state organizations. UAVs, including relatively simple and inexpensive models, can be used to collect and transmit targeting data on Navy ships, attack Navy ships directly by diving into them, and be armed to attack Navy ships at a distance. ASCMs are widely proliferated to state actors, and were also reportedly used by the non-state Hezbollah organization in 2006 to attack an Israeli warship. China has developed an ASBM. Lasers that are not capable of disabling ballistic missiles could nevertheless augment ballistic missile defense operations by being used for precision tracking and imaging.

Required Laser Power Levels for Countering Targets

A laser's ability to disable a target depends in large part on the power and beam quality of its light beam. The power of the light beam is measured in kilowatts (kW) or megawatts (MW). Beam quality (BQ) is a measure of how well focused the beam is.[15] Additional factors affecting a laser's ability to disable a target include:

- atmospheric absorption, scattering, and turbulence,[16]
- jitter—the degree to which the spot of laser light jumps around on the surface of the target due to vibration or other movement of the laser system,[17] and
- target design features, which can affect a target's susceptibility to laser damage.

Table A-1 in Appendix A summarizes some government and industry perspectives regarding power levels needed to counter certain targets. Although these perspectives differ somewhat, the following conclusions might be drawn from the table regarding approximate laser power levels needed to affect certain targets:

- **Lasers with a power level of about 10 kW** might be able to counter some UAVs at short range, particularly "soft" UAVs (i.e., those with design features that make them particularly susceptible to laser damage).
- **Lasers with power levels in the tens of kilowatts** could have more capability for countering UAVs, and could counter at least some small boats as well.

- **Lasers with a power level of about 100 kW** would have a greater ability for countering UAVs and small boats, as well as some capability for countering rockets, artillery, and mortars.
- **Lasers with power levels in the hundreds of kilowatts** could have greater ability for countering targets mentioned above, and could also counter manned aircraft and some missiles.
- **Lasers with power levels in the megawatts** could have greater ability for countering targets mentioned above—including supersonic ASCMs and ballistic missiles—at ranges of up to about 10 nautical miles.

In addition to the points above, one Navy briefing stated that lasers with power levels above 300 kW could permit a ship to defend not only itself, but other ships in the area as well (a capability referred to as area defense or escort operations or battle group operations).

Types of Lasers Being Developed for Potential Shipboard Use

The Navy and DOD are developing three principal types of lasers for potential use on Navy surface ships:

- fiber solid state lasers (SSLs),
- slab SSLs, and
- free electron lasers (FELs).

All three types are electrically powered.[18] Each type is discussed briefly below. Additional information on each type is presented in **Appendix C** through **Appendix F**.

Fiber Solid State Lasers (Fiber SSLs)

Fiber solid state lasers (SSLs) are widely used in industry—tens of thousands are used by auto and truck manufacturing firms for cutting and welding metal. Consequently, they are considered to be a very robust technology.

Laser Weapon System (LaWS)

One fiber SSL prototype demonstrator developed by the Navy, called the **Laser Weapon System (LaWS)**, had a beam power of 33 kW. The Navy at one point envisioned LaWS being used for operations such as disabling or reversibly jamming EO sensors, countering UAVs and EO guided missiles, and augmenting radar tracking. The Navy envisioned installing LaWS on a ship either on its own mount or as an add-on to an existing Phalanx Close-In Weapon System (CIWS) mount.[19] The Navy funded work to integrate LaWS with CIWS, to support the latter option.

The Navy has stated the following regarding tests of LaWS:

- In June 2009, LaWS successfully engaged five threat-representative UAVs[20] in five attempts in tests in combat-representative scenarios in a desert setting at the Naval Air Weapons Station at China Lake, in southern California.
- In May 2010, LaWS successfully engaged four threat-representative UAVs in four attempts in combat-representative scenarios at a range of about one nautical mile in an over-the-water setting conducted from San Nicholas Island, off the coast of southern California. LaWS during these tests also demonstrated an ability to destroy materials used in rigid-hull inflatable boats (RHIBs—a type of small boat) at a range of about half a nautical mile, and to reversibly jam and disrupt electro-optical/infrared sensors.[21]
- Between July and September 2012, LaWS successfully engaged three UAVs in three attempts in tests during which the system was aboard the Arleigh Burke (DDG-51) class destroyer *Dewey* (DDG-105) in waters off San Diego.[22]

The Navy at one point envisioned scaling up the power of the LaWS beam to about 100 kW by FY2014. How much beyond 100 kW the system could eventually be scaled up to was not clear, but the system was not generally viewed as having the potential for being scaled up to megawatt power levels.

The Navy stated that as of June 2010, the Technology Readiness Level (TRL) of the LaWS prototype "is approaching 6, based on a system prototype demonstration in a relevant (maritime) environment."[23] The Navy estimated that it might cost roughly $150 million to develop LaWS to TRL 7, meaning the demonstration of a system prototype in an operational environment. The Navy considered the LaWS effort to be ready for conversion into a program of record. The Navy estimated that production copies of the LaWS system could be installed and procured as additions to ship CIWS mounts for a total cost of roughly $17 million per CIWS mount.[24]

The Navy in August 2014 installed a LaWS system on the USS *Ponce*, a ship operating in the Persian Gulf as an interim Afloat Forward Staging Base (AFSB[I]), to conduct continued evaluation of shipboard lasers in an operational setting. For further discussion, see "LaWS Installed and Tested on USS Ponce" in "Recent Developments" below.

For additional information on fiber SSLs and LaWS, see **Appendix C**.

Tactical Laser System

Another Navy fiber SSL effort is the **Tactical Laser System (TLS)**—a laser with a beam power of 10 kW that is designed to be added to the Mk 38 25 mm machine guns installed on the decks of many Navy surface ships.[25] TLS would augment the Mk 38 machine gun in countering targets such as small boats; it could also assist in providing precise tracking of targets. The Navy in March 2011 awarded a $2.8 million contract to BAE to develop a prototype of the TLS over a 15- month period.[26] Boeing is collaborating with BAE on the project. The TLS effort was initiated following a January 2008 incident involving Iranian small boats.

A March 26, 2012, press report states that "[Michael] Rinn, [Boeing's vice president for directed energy systems], said the project, which gets a small amount of Navy funding and is supplemented by internal investments from both companies, has had several successes over the past few years. Field testing of the major components last summer at Eglin Air Force Base in Florida showed the system could distinguish between friendly and enemy activities in both daytime and nighttime, for example." The report states that full system testing of the laser was scheduled for the summer of 2012.[27]

A January 28, 2013, press report states that "BAE Systems and Boeing are pushing their Mark 38 Tactical Laser System through intense internally-funded testing, hoping to get onto a ship for at-sea testing as soon as one is available, company officials said earlier this month." The article stated that Mark Rinn, a Boeing official, "said testing in December [2012] went well and showed successful engagements at 'several thousands of meters.' He had hoped to include unmanned aerial vehicles in the set of targets for the Mark 38 Tactical Laser System—the weapon system has already shot at targets on land and on water—but the companies could not get permission in time. He said that they would have permission for counter-UAV testing before the next round of tests this spring."[28]

For additional information on TLS, see **Appendix D**.

Slab Solid State Lasers (Slab SSLs)

DOD has pursued multiple efforts to develop slab SSLs for military use. Among these was the **Maritime Laser Demonstration (MLD)**, a prototype laser weapon developed as a rapid demonstration project under DOD's Joint High Power SSL (JHPSSL) program. MLD leveraged development work on slab SSLs done elsewhere in DOD under the JHPSSL program. In March 2009, Northrop demonstrated a version of MLD that coherently combined seven slab SSLs, each with a power of about 15 kW, to create a beam with a power of about 105 kW.

In July 2010, the ability of MLD to track small boats in a marine environment was tested at NSWC Port Hueneme, CA.[29] In late August and early September 2010, MLD was tested in an over-the-water setting at the Navy's Potomac River Test Range against stationary targets, including representative small boat sections.[30] In November 2010, an at-sea test of the system against small boat targets reportedly was stopped midway because one of the system's components needed to be replaced.[31] The test was resumed in April 2011, and on April 6, 2011, the system successfully engaged a small target vessel. According to the Navy, this was the first time that a laser of that energy level had been put on a Navy ship, powered from that ship, and used to counter a target at range in a maritime environment.[32] In May 2011, Northrop stated that it could build the first unit of a full-power engineering and manufacturing development (EMD) version of the weapon within four years, if the Navy could find the resources to fund the effort.[33]

Scaling up a slab laser to a total power of 300 kW is not considered to require any technological breakthroughs. Supporters of slab SSLs such as MLD believe they could eventually be scaled up further, to perhaps 600 kW. Slab SSLs are not generally viewed as easily scalable to megawatt power levels.

The Navy stated that as of December 2010, MLD was at a Technology Readiness Level (TRL) of 5, meaning component and/or breadboard validation in a relevant environment.[34]

For additional information on slab SSLs and MLD, see **Appendix E**.

Free Electron Lasers (FELs)

Unlike slab SSLs, which are being developed by multiple U.S. military services, FELs are being developed within DOD solely by the Navy, in part because they would be too large to be installed on Army or Marine Corps ground vehicles or Air Force tactical aircraft, and in part because an FEL's ability to change its wavelength so as to match atmospheric transmission sweet spots makes it particularly suited for operations in a marine environment. The basic architecture of an FEL offers a clear potential for scaling up to power levels of one or more megawatts.

A 14.7 kW FEL has been developed; it has not been moved out of a laboratory setting or fired at an operational moving target. The Office of Naval Research (ONR) had planned to follow this with the development, as an Innovative Naval Prototype (INP),[35] of a 100 kW FEL; the work was scheduled to be performed during FY2010-FY2015.[36] Developing a 100 kW FEL would reduce the risks associated with developing a megawatt-class FEL. A March 26, 2011, press report, however, states that "the Navy is putting the project on the back burner as it focuses on a solid-state laser as the quickest way to get a directed-energy weapon to the fleet." The report states that "[Roger] McGinnis, [program executive for INPs at ONR's Naval Air Warfare and Weapons Department], said the Navy had previously wanted to pursue a 100 kilowatt FEL gun as an intermediate step toward the megawatt gun but decided to instead focus on maturing the critical technology components with an Energy department lab or small industry partners.... "[37]

The Navy states that as of December 2010, FEL was at a Technology Readiness Level (TRL) of 4 (meaning component and/or breadboard validation in a laboratory environment).[38]

For additional information on FEL, see **Appendix F**.

Navy Surface Fleet's Generalized Vision for Shipboard Lasers

The Navy's surface fleet has a three-phase generalized vision for shipboard high-energy lasers that is summarized in **Table 1**. Although this generalized vision refers to lasers of certain power levels and potential time frames for installing lasers on Navy ships, it is not a program of record for procuring a production version of a shipboard laser.

Remaining Technical Challenges

Although Navy and DOD research on military lasers has overcome many of the technical challenges associated with developing shipboard lasers, a number of challenges remain. Remaining technical challenges for potential shipboard lasers can be grouped into four broad categories:

- scaling up beam power to higher levels while maintaining or improving beam quality and addressing thermal management (the removal of waste heat from the gain medium);
- turning prototype and demonstration versions of lasers into versions that are suitable for series production, shipboard installation, and shipboard operation and maintenance over many years of use;
- engineering other parts of a complete laser weapon system, including target detection and tracking, and beam pointing; and
- integrating lasers with ship power and cooling systems, and with ship combat systems (i.e., a ship's integrated collection of sensors, computers, displays, and weapons).

Although these challenges are stated briefly here, they are not trivial. Skeptics might argue that certain past DOD laser development efforts proved over-optimistic in terms of projections for overcoming technical challenges and producing operational weapons. In spite of decades of development work, these skeptics might note, DOD has not deployed an operational high-energy laser weapon system.

Table 1. Surface Navy's Generalized Vision for Shipboard High-Energy Lasers (Draft version as of May 2011)

Initial capability		Added capability	Added capability
Laser's beam power	60 kW to 100 kW	300 kW to 500 kW	> 1 MW
Missions	Countering UAVs, EO-guided ASCMs, enemy ISR systems, and swarm boats, and used for precise tracking to support air defense missions conducted by electromagnetic rail gun (EMRG), ballistic missile defense (BMD) missions, augmenting the ship's radar, and enhancing general situational awareness	Capabilities in previous column, but with added range and a capability to counter ASCMs flying a crossing path toward another ship.	Capabilities in previous column, but a capability for full-self defense operations against ASCMs and maneuvering reentry vehicles (MaRVs), and full MD missions

Table 1. (Continued)

Initial capability		Added capability	Added capability
Required ship power (in kW or MW) and cooling capacity (in tons)a	< 400 kW and 68 tons	< 2.5MW and 560 tons	~ 10-20 MW and ~ 1,400 tons
Current weapon system TRL	5	4	2-3
Earliest potential IOC	2017	~ 2022	after 2025
Applicable ships	Could be backfit onto existing ships, as well as installed on new ships	Could be installed on future surface combatants, including potentially the Flight III DDG-51	Could be installed on future surface combatants, ships with integrated propulsion systems, and aircraft carriers

Source: U.S. Navy briefing slide dated May 20, 2011, and provided to CRS and CBO at a briefing on that date.

[a] Power and cooling requirements assume continuous firing of the laser with a 67% duty cycle (i.e., the laser is firing 67% of the time).

Recent Developments

Naval Directed Energy Steering Group

In June 2012, it was reported that the Navy in December 2011 formed a Naval Directed Energy Steering Group (NDESG) to develop a naval directed energy vision, strategy, and roadmap. The December 12, 2011, Navy memorandum establishing the steering group states in part:

> A key to future Navy and Marine Corps war fighting capabilities is the efficient, effective and rapid development, acquisition, and fielding of advanced technologies having game-changing capabilities across a range of mission areas. Directed Energy Weapon (DEW) technologies, including lasers and high power microwave (HPM) weapons, may offer our naval forces such game-changing potential....

The Naval DE Steering Group (NDESG) is formed as a Secretary of the Navy (SECNAV) initiative to deliver a synchronized, fiscally-informed strategy that aligns DE investments with roadmaps across the Doctrine, Organization, Training, Material, Leadership and Education, Personnel and Facilities (DOTMLPF) spectrum [of naval activities] to address near-term fleet capability gaps and the long-range vision for the implementation of DE in the fleet. The NDESG will be the formal engine to drive this effort....

The NDESG will have the following objectives:

a. Develop a DON Naval DE Vision and Strategy.... A Directed Energy vision is necessary to provide DON leadership's depiction of desired DEW capabilities and DE countermeasures as deployed and employed across U.S. naval forces. A supporting DE strategy would be used to establish strategic goals, guiding principles, mission area priorities, roles and responsibilities and overarching objectives regarding the acquisition and fielding of DEW across the Navy and Marine Corps.

b. Develop a comprehensive DE roadmap... based on the overarching vision and strategy. The proposed roadmap would address the prioritized mission needs across all naval forces and the associated DE technologies than can be fielded to satisfy those mission needs in the near-term (2-5) years, mid-term (5-10 years) and far-term (10-20 years).

c.Provide assessments on Science & Technology (S&T)/Research & Development (R&D) and oversee the development and transition of DE systems and technologies to the Fleet, including non-material efforts[39] to integrate these new capabilities into existing operational concepts and procedures....

The NDESG will provide a draft vision and strategy with initial plan of actions and milestones to the UNDERSECNAV [Under Secretary of the Navy] within 90 days of the promulgation of this charter.[40]

Directed Energy Vision for U.S. Naval Forces

The directed energy vision and the directed energy strategy called for in paragraph (a) of the memorandum quoted in the previous section have been developed. The text of the vision statement is as follows:

A Directed Energy Vision for U.S. Naval Forces

Guidance from the Secretary of Defense promulgated in Priorities for 21st Century Defense in January 2012 directs the Department to "sustain key streams of innovation that may provide significant long-term payoffs." Directed-energy (DE) technology not only offers the prospect for a major return on investment over the long term, it could begin paying significant dividends within the current future years defense plan (FYDP) by addressing immediate combatant commander requirements and enabling fleet experimentation focused on emerging threats, including anti-access and area-denial challenges.

Military applications of DE technology hold growing promise for gaining and sustaining tactical, operational, and strategic advantage for U.S. forces across the full range of military operations. They could have significant effects across multiple dimensions of the battlespace: maritime, air, land, space, and cyberspace. Directed energy weapons (DEWs) offer several potentially "game changing" advantages: very rapid engagement, low cost per engagement, essentially infinite magazines, and low total ownership costs. DEWs and their associated platform integration technologies must be properly resourced across the FYDP to ensure that our Navy and Marine Corps Team maintains its warfighting edge over prospective adversaries, including those aggressively pursuing DEWs.

DEWs affect a target by imparting non-kinetic, or electromagnetic, energy. DEW technologies can operate in any part of the electromagnetic spectrum and typically fall into the categories of either lasers (i.e., low, medium, or high power) or high-power radio frequency (i.e., high-power microwave, radio frequency (RF), microwave, and millimeter wave (MMW)). DEW technologies and systems use electromagnetic energy to cause persistent disruption, reversible effects or permanent damage by attacking target materials, electronics, optics, antennas, and sensors, including non-lethal counter-personnel and counter-materiel applications. The ability of these weapons to incapacitate, disrupt, damage, disable, or destroy targets has been proven with numerous demonstrations of lethal and nonlethal effects carried out in laboratory, field testing and evaluation, and successful employment on the battlefield.

The DoN [Department of the Navy] will focus its DE investments on those technologies that address critical Navy and Marine Corps capability gaps. Given the surface fleet's ability to overcome the technical challenges associated with the military exploitation of high power, long range DEW—including power, cooling, weight, and volume requirements—it is the logical vanguard for demonstrating the potential of first-generation weapons. Across the spectrum of DEWs, early applications will focus on supporting forward deployed forces to defeat Improvised Explosive Devices (IEDs); artillery, mortars, and rockets; intelligence, surveillance and reconnaissance systems; fast-attack craft; fixed and rotary-wing aviation; and subsonic anti-ship cruise missiles. The longer term objective is to field higher power systems capable of defeating supersonic cruise missiles and selected ballistic missiles.

As the technology matures to increase energy efficiency and reduce form factors, DEWs will be integrated into ground vehicles to support fire and maneuver in contested environments, to include conducting low-collateral damage strikes in built-up terrain, employing non-lethal DEW to segregate and isolate enemy from civilians, and defending against increasingly ubiquitous guided rockets, artillery, mortars, and missiles. DE applications for fixed- and rotary-wing aircraft will focus both on offensive and defensive air-to-air, air-to-surface, and air-to-ground missions. Early applications will focus on countering surface-to-air and small boat threats, as well as conducting precision strikes with mission-tailored lethality.

The DoN will field initial DEW capabilities in the near-term to provide our fleet and operating forces with the ability to address identified critical mission capability gaps while learning invaluable fielding and employment lessons that will inform our way ahead. Innovation has been the hallmark of U.S. Naval Forces. DEWs represent another naval innovation that when transitioned from the laboratory to battlefield will help our Navy and Marine Corps Team to sustain its technological advantage and win our nation's battles. Towards this end, the DoN will take a measured approach toward DEW S&T and R&D activities and their transition to acquisition programs based on operational requirements, technological maturity or readiness, demonstrated performance, ease of systems integration and affordability.

The DoN will address the defensive challenges posed by diffusion and maturation of DEWs available to prospective adversaries. These efforts will guide the development and fielding of countermeasures, DEW-resistant systems, and effective non-material solutions across the maritime battlespace domain. While high-power DEWs will be limited to nation states that choose to pursue them, lower power weapons will become increasingly available at a relatively low cost to non-state actors.

Finally, the DoN will coordinate with other Services and agencies to ensure policies and rules of engagement are in place to enable our Sailors and Marines to operationally employ DEWs effectively. In addition, we will develop not only the DEWs themselves but the sensors, communications, and control technologies that will enable DEWs to operate, in combination with other military capabilities, at their full potential.[41]

Directed Energy Roadmap and Possible Analysis of Alternatives (AOA)

An August 5, 2013, press report based on an interview with a Navy official states that the Naval Directed Energy Steering Group "will have its near-term roadmap ready this fall to begin informing decisions to address drone and small boat swarm threats with directed-energy weapons rather than kinetic weapons, with mid- and long-term roadmaps to follow next year." The report quoted the Navy official as saying that there have been discussions of conducting an analysis of alternatives (AOA) on directed-energy capabilities in FY2014.[42]

Destroyers and LCSs Reportedly Leading Candidate Platforms

An August 20, 2012, press report stated that following the MLD effort, the Navy conducted studies to examine the ability of various Navy ship classes to accept SSLs. The report quoted Peter Morrison, ONR's SSL program manager, as saying that based on these studies, "the DDG [destroyer] and LCS [Littoral Combat Ship] classes ... provided the best opportunity to match new capabilities with emerging needs with higher-energy laser weapons capabilities, and the class' forecasts for power, cooling, space and weight." The report stated that the Navy continues to review the potential for installing SSLs on other types of ships as well.[43]

LaWS Installed and Tested on USS Ponce, Declared Operational

On April 8, 2013, the Navy announced that it would install LaWS on the USS *Ponce* (pronounced pon-SAY), a converted amphibious ship that is operating in the Persian Gulf as an interim Afloat Forward Staging Base (AFSB[I]), to conduct evaluation of shipboard lasers in an operational setting against swarming boats and swarming UAVs.[44] LaWS, the Navy stated, would be installed on the *Ponce* in the summer of 2014 and would be evaluated on the ship for a period of 12 months.[45]

The system was installed on the *Ponce* in August 2014. A December 10, 2014, news release from the Office of Naval Research (ONR) stated:

> Officials at the Office of Naval Research (ONR) announced today that the laser weapon system (LaWS)—a cutting-edge weapon that brings significant new capabilities to America's Sailors and Marines—was for the first time successfully deployed and operated aboard a naval vessel in the Persian Gulf.
>
> The operational demonstrations, which took place from September to November [2014] aboard USS Ponce (AFSB[I] 15), were historic not only because they showed a laser weapon working aboard a deployed U.S. Navy ship, but also because LaWS operated seamlessly with existing ship defense systems....
>
> "Laser weapons are powerful, affordable and will play a vital role in the future of naval combat operations," said Rear Adm. Matthew L. Klunder, chief of naval research. "We ran this particular weapon, a prototype, through some extremely tough paces, and it locked on and destroyed the targets we designated with near-instantaneous lethality."
>
> During the tests, LaWS—a collaborative effort between ONR, Naval Sea Systems Command, Naval Research Laboratory, Naval Surface Warfare Center Dahlgren Division and industry partners—hit targets mounted aboard a speeding oncoming small boat, shot a Scan Eagle

unmanned aerial vehicle (UAV) out of the sky, and destroyed other moving targets at sea.

Sailors worked daily with LaWS over several months since it was installed, and reported the weapon performed flawlessly, including in adverse weather conditions of high winds, heat and humidity. They noted the system exceeded expectations for both reliability and maintainability.

The system is operated by a video-game like controller, and can address multiple threats using a range of escalating options, from non-lethal measures such as optical "dazzling" and disabling, to lethal destruction if necessary. It could prove to be a pivotal asset against what are termed "asymmetric threats," which include small attack boats and UAVs.

Data regarding accuracy, lethality and other factors from the Ponce deployment will guide the development of weapons under ONR's Solid-State Laser-Technology Maturation program. Under this program, industry teams have been selected to develop cost-effective, combat-ready laser prototypes that could be installed on vessels such as guided-missile destroyers and the Littoral Combat Ship in the early 2020s....

"At less than a dollar per shot, there's no question about the value LaWS provides," said Klunder. "With affordability a serious concern for our defense budgets, this will more effectively manage resources to ensure our Sailors and Marines are never in a fair fight."

The Navy already has demonstrated the effectiveness of lasers in a variety of maritime settings. In a 2011 demonstration, a laser was used to defeat multiple small boat threats from a destroyer. In 2012, LaWS downed several unmanned aircraft in tests during naval exercises. Specific details on next steps and timeframes are being determined as the data from the current demonstrations are analyzed.[46]

A December 11, 2014, trade press report stated:

The Navy's first-of-a-kind laser deployed on a vessel sailing in the Persian Gulf has been declared operational and can be used by the crew to defend itself against potential threats, the service's head of the Office of Naval Research said on Wednesday.

Rear Adm. Matthew Klunder told reporters on a conference call that Central Command has been green lighted to use the laser in the event of a threat, approval that has been passed along to the ship's commanding officer. The 30-kilowat laser, known as the Laser Weapon System, or LaWS, was installed on the USS Ponce in August.

The ship later departed for the Persian Gulf and the LaWS successfully carried out operational testing recently by striking a fast attack boat and drone, Klunder said, adding that this marks the "historic" first ever operational deployment of a directed energy weapon.[47]

Navy Anticipates Program of Record in FY2018 and IOC in FY2020-FY2021

In March 2014, it was reported that the Navy anticipates moving to a shipboard laser program of record in "the FY2018 time frame" and achieving an initial operational capability (IOC) with a shipboard laser in FY2020 or FY2021.[48]

March 2014 Navy Testimony

At a March 26, 2014, hearing before the Intelligence, Emerging Threats & Capabilities subcommittee of the House Armed Services Committee on FY2015 DOD Science and technology Program, Rear Admiral Matthew L. Klunder, Chief of Naval Research, stated:

> An ongoing example of our success is the laser weapons system [LaWS], part of our solid state laser maturation effort [SSL-TM]. We feel energy weapons, specifically directed energy weapons, offer the Navy and the Marine Corps game-changing capabilities in speed of light engagement, deep magazines, multi-mission functionality and affordable solutions. Laser weapons are very low engagement costs. Right now, we're literally under a U.S. dollar per— per pulsed energy round. Which is critical in our current fiscal environment.
>
> They are capable in defeating adversarial threats, including fast boats, UAVs and other low-cost, widely-available weapons. Now, our laser weapons system—again, referred to as LaWS—leverages advances in commercial technology for use in a rugged, robust prototype weapon capable of identifying, illuminating, tracking and lasing enemy surface and air threats. The Navy's installing this LaWS system on board the USS Ponce in the Arabian Gulf this year; this summer, to be exact.
>
> That harsh and operationally important environment will provide an ideal opportunity to evaluate long-term system performance. We believe that LaWS has every potential for extraordinary success in field—terms of fielding an effective, affordable weapon for our sailors and Marines.[49]

Later in the hearing, the following exchange occurred:

> REPRESENTATIVE NUGENT (continuing): ... Admiral, I'm really interested in—and I'm interested in all of you as it relates to directed energy. Mr. Langevin and I, I think, are—are pretty big proponents of directed energy because of what you mentioned in regards to—on the Ponce, in regards to actually testing, and the ability to test and what it costs to test versus shooting a missile off at a—a million dollars a copy versus a dollar.

Can you—we see programs in development stage. But then they tend to never make it to production, never make it to, you know, deployment. Where do we stand as it relates to that system on the Ponce in regards to the future?

KLUNDER: Yes, sir. Thank you for the question. And I—I'll offer that there's—it's really a conviction my our senior leadership in the Department of the Navy. And—and what I mean by that is that we—we want those new innovative systems to be in the hands of sailors and Marines. We want them to tell us did we develop it right, did we develop and it needs to be tweaked a little bit? Or did we develop and we just didn't do it right? And we'll—we'll bring it back.

But the point there is, you need to get a sailor or a Marine's hands on that thing, and tell them is it gonna be effective in warfighting environment, and it—will—will it be affordable. So the point I'd like to make, and thank you for your—your—your comments about innovation, we truly think that's the way this nation was built and—and is the way we get in front of our adversaries. We don't want to run with them. I don't want a sailor or a Marine to ever go into a fair fight.

I want them to always have the technological advantage so we always win and defend our nation. What we've done this time on the Ponce, I think, is very credible is I don't have a bunch of (ph)—my scientists and my colleagues, we developed it. But I've got real sailors right down there at Dahlgren, right now, on the system. And it's not a singular laptop over in the corner somewhere. It's a fully-integrated with our fully-integrated combat information system on that ship.

So those young men and women on that—detachment of sailors are gonna go out there. They're gonna test it. And, indeed, we feel very comfortable because we've never missed so far. And that's one of the reasons why CNO [the Chief of Naval Operations, Admiral Jonathan] Greenert said, "Matt, get it out there." We've never missed. We feel confident, though, that we'd like to test it in that tough environment and see where it goes.

And to—the follow-on to the last bit of your question, I think regardless of the High Energy Laser-Joint Technology Office, I can assure you that we've got all the resources positioned in the Navy and Marine Corps to put us in a good place when this test is done. And I'm not sure if you're familiar, but we also have a solid state laser technology maturation program that takes it to a much higher power level, and that's in '16 [FY2016].

So when we finish this test on Ponce, that demo with real sailors, and we finish up the prototyping in '16 [FY2016], we think we'll be very well positioned for follow-on, longterm, enduring efforts.

NUGENT: And I just don't want us to—we can be in a testing mode forever. KLUNDER: Yes, sir.

NUGENT: I mean, I think you might agree with that. And—and I'd like to see us have at least a timeline as to—as to when we want to have it operational. It goes back to CHAMP.[50] Mr. Langevin and I have talked about that. It goes back to programs as it relates to the Army, and I know there's some collaboration between the Army and the Navy on those issues. And from my standpoint, I think that's great when you can get bright minds across the lines, across those services, to utilize that same information and—and make us all safer.

So my question back to you, then is, if, after this test on the Ponce, if it meets the expectations, what would stand in your way of, if it's successful, in deploying that on other ships?

KLUNDER: I would say nothing. Right now, we've already started the AOA on that process, that we're very familiar with the acquisition programs. We've already done all the blueprinting for the different classes of ships. So in many cases, if we are successful we see this as a possible weapons system for a number of classes of our ships. And I think it's important, too, if I could just give my colleagues to my right here a great shout out.

Because we're doing a test down in your—your great state here in just a few months here to do some joint Army-Navy testing down at Eglin. And so I think that, again, shows the collaborative effort we do on directed energy.[51]

FY2016 Funding Request

The Navy's proposed FY2016 budget requests $26.9 million for research and development work on directed energy technologies, including the SSL technologies, in the directed energy portion of Program Element (PE) 0602114N, Power Projection Applied Research, a line item in the Navy's research and development account.[52] The figures for FY2014 and FY2015 for this PE are $40.3 million and $40.5 million, respectively. The Navy states that the "FY 2015 to FY 2016 decrease in funding is due to completion of the Solid State Laser - QRC [Quick Reaction Capability] program as well as a continued realignment of FEL activities."[53]

Additional funding for the Solid State Laser Technology Maturation Program (SSL-TM) forms part of the FY2016 funding request for the precision strike technology portion of PE 0603114N, Power Projection Advanced Technology, another line item in the Navy's research and development account.[54]

The Navy's proposed FY2016 budget also requests $9.5 million for SSLs in Project 9823 (Lasers for Navy Applications) within PE 0603925N, Directed

Energy and Electric Weapon System, another line item in the Navy's research and development account.[55]

ISSUES FOR CONGRESS

Number of Laser Types to Continue Developing

Potential Strategies

One potential issue for Congress is how many of the three laser types discussed in this report— fiber SSLs, slab SSLs, and FELs—the Navy should continue developing.

Supporters of stopping development of all three types (or of continuing development of one type) might argue that continuing the development of shipboard lasers (or of more than one type of laser), while perhaps desirable, would reduce funding for more important Navy program priorities below critical levels, particularly in a situation of constrained Navy resources. They might argue that the Navy's kinetic weapons in coming years will have sufficient (or largely sufficient) capability for countering the kinds of targets that shipboard lasers could counter.

Supporters of continuing development of two or three types might argue that it would permit continued competition between laser types and provide a hedge against the failure of one of the development efforts. DOD in the past, they might argue, has sometimes pursued comparable programs concurrently to ensure the best outcome for an area of effort deemed important. They might also argue that the Navy's kinetic weapons in coming years will be insufficient to counter certain kinds of targets, or that shipboard lasers would counter them more cost effectively.

Relative Merits of Laser Types

In considering which laser types to continue developing, policy makers may consider the relative merits of each type. Below are some arguments relating to the relative merits each type. The discussions below are intended as introductory only; a full comparison of their relative merits would entail much longer discussions.

Some Arguments Relating to Fiber SSLs

Supporters of LaWS argue that it has a demonstrated ability to counter certain targets of interest at short (but tactically useful) ranges in a marine environment; that it can be installed on Navy ships in the near term; that it promises to be less expensive than a slab SSL; that it poses less of a challenge in terms of thermal management than a slab SSL; that it has less ship impact than FELs; that it uses an industrial laser technology with high reliability and few alignment optics, making possible a simplified system engineering solution for a Navy laser system; and that its power can be scaled up to 100 kW or perhaps more. They argue that the system's BQ, though not excellent, is good enough to disable targets of interest at short ranges. They argue that the system's light wavelength of 1.064 microns, though not exactly on the atmospheric transmission "sweet spot" located at 1.045 microns, is good enough in terms of atmospheric transmission to permit the laser to disable targets of interest at tactically useful ranges, and that development work is underway on SSLs that would emit light at wavelengths above the threshold (about 1.5 microns) at which laser light becomes much less dangerous to human eyes.

Some skeptics of LaWS, including supporters of the MLD, argue that the LaWS's BQ limits its effective range. Other skeptics of LaWS, including supporters of FELs, argue that LaWS's operating wavelength limits its effective range, particularly when compared to FELs, whose wavelengths can be tuned to exactly match atmospheric transmission sweet spots, and that LaWS's current wavelength is dangerous to human eyes, whereas an FEL can operate at wavelengths matching atmospheric sweet spots that are located above 1.5 microns.

Some Arguments Relating to Slab SSLs

Supporters of MLD argue that it has a demonstrated power level of 105 kW (more than three times that of LaWS); that it has a much better BQ than LaWS, permitting it to counter targets at greater ranges (thereby providing a larger defended area around the ship, and more time to counter targets approaching the ship); that it could be ready for installation on ships as soon as, or not very long after, the LaWS system would be; that a production version could have a procurement cost comparable to, or even less than, that of a production version of LaWS; that the challenge slab SSLs pose in terms of thermal management, though perhaps greater than that of fiber SSLs, can nevertheless be handled; and that slab SSLs can be scaled up to 300 kW or more while retaining good BQ. The MLD contract, they argue, was competitively awarded, the competitors for the contract included fiber SSLs, and the contract was awarded instead to a slab SSL.

Supporters of slab MLDs argue that the difference in complexity between fiber SSLs and slab SSLs is not as great as some supporters of LaWS contend—that fiber SSLs, for example, have more free-space optics[56] than slab SSLs. Supporters of MLD argue that the industrial environments in which commercial fiber SSLs have operated are not characterized by shocks or high humidity—two features that characterize the shipboard operating environment—whereas MLD was designed from the start with eventual ship operations in mind. Supporters of MLD argue that it can be maintained easily in the field through the use of sealed line replaceable units (LRUs).[57] MLD supporters argue, as do supporters of LaWS, that the system has less ship impact than an FEL; that the system's light wavelength of 1.064 microns, though not exactly on the atmospheric transmission "sweet spot" located at 1.045 microns, is good enough in terms of atmospheric transmission to permit the laser to disable targets of interest at tactically useful ranges, and that development work is underway on SSLs that would emit light at wavelengths above the threshold (about 1.5 microns) at which laser light becomes much less dangerous to human eyes.

Skeptics of MLD, including supporters of LaWS, argue that it uses complex optics, making it more expensive to procure and potentially less reliable and more difficult to maintain than LaWS. Other skeptics of MLD, including supporters of FELs, argue, as they do regarding LaWS, that MLD's operating wavelength limits its effective range, particularly when compared to FELs, whose wavelengths can be tuned to exactly match atmospheric transmission sweet spots, and that MLD's current wavelength is dangerous to human eyes, whereas an FEL can operate at wavelengths matching atmospheric sweet spots that are located above 1.5 microns.

Some Arguments Relating to FELs

Supporters of FELs argue that unlike SSLs, FELs clearly can be scaled up to megawatt power levels that would be capable of countering a wide range of targets, including supersonic ASCMs and ballistic missiles, and that unlike SSLs, FELs can be scaled up in power from 10 kW to 1 MW without any increase in the size of the system or need for a beam combiner (a component that adds to system complexity and cost). Supporters of FELs argue that in contrast to the fixed wavelength of light emitted by an SSL, the wavelength of light emitted by an FEL can be tuned to exactly match various atmospheric transmission sweet spots, including those above the threshold (about 1.5 microns) at which laser light becomes much less dangerous to human eyes. They also argue that in contrast to SSLs, FELs pose no large thermal management issues because an FEL's waste heat is not produced inside the laser mechanism itself.

Skeptics of FELs, including supporters of SSLs, argue that FELs will not be ready for installation on ships for a significant number of years. They argue that FELs are so large that they cannot be incorporated into most if not all existing Navy ship designs, limiting the potential applicability of FELs to the surface fleet for many years, and that incorporating an FEL into a new ship design could make the ship considerably larger, adding to the ship's construction cost. They also argue that the need for isolating the FEL system from vibration and shock and the possible need for using cryogenic equipment adds to an FEL's cost and complexity.

Implications for Ship Design and Acquisition

Another potential issue for Congress are the possible implications that shipboard lasers might have for the design and acquisition of Navy ships, including the Flight III DDG-51 destroyer that the Navy wants to begin procuring in FY2016.[58] The ability of existing Navy ship designs to support lasers, particularly in terms of having sufficient electrical power and cooling capacity, can be summarized as follows:

- The Navy has concluded that its Aegis cruisers and destroyers (i.e., CG-47 and DDG-51 class ships), as well as San Antonio (LPD-17) class amphibious ships, would have enough available electrical power under battle conditions (i.e., when many other systems are also drawing electrical power) to support a LaWS system. An August 2010 press report stated: "Today's warships have enough power to support a 100-kilowatt laser, said [Capt. David Kiel, program manager for directed energy and electric weapons at Naval Sea Systems Command]. Any surface combatant large enough to accommodate the close-in weapon system [CIWS] could also carry the fiber laser, he added."[59]
- Some Navy ships might be able to support, under battle conditions, an SSL with a power *somewhat* above 100 kW.
- No existing Navy surface combatant designs have enough electrical power or cooling capacity to support an SSL with a power level *well* above 100 kW.

- Because of its probable size, an FEL could not be backfitted onto existing cruisers or destroyers. Aircraft carriers and "large-deck" amphibious assault ships (i.e., LHA/LHD-type amphibious ships) might have enough room to accommodate an FEL, but existing carriers and amphibious assault ships might not have enough electrical power to support a megawatt-class FEL. In addition, because of thermal blooming and the status of carriers and amphibious assault ships as potential high-value targets, it might make more operational sense to install megawatt-class FELs on ships other than carriers or amphibious assault ships.[60]

The above points suggest that the Navy in coming years could face significant ship-design constraints in its ability to install shipboard lasers, particularly SSLs well above 100 kW in power, and FELs in general. These constraints are a product, in part, of the Navy's termination of the CG(X) cruiser program, because the CG(X) could have been designed to support SSLs well above 100 kW in power and/or a megawatt-class FEL.[61] Following the termination of the CG(X) program, the Navy has no announced plans to acquire a surface combatant clearly capable of supporting an SSL well above 100 kW in power, or an FEL.

Ship-design options for expanding the Navy's ability to install lasers on its surface ships in coming years include the following:

- design the new Flight III version of the DDG-51 destroyer, which the Navy wants to start procuring in FY2016, with enough space, electrical power, and cooling capacity to support an SSL with a power level of 200 kW or 300 kW or more—something that could require lengthening the DDG-51 hull, so as to provide room for laser equipment and additional electrical generating and cooling equipment;
- design and procure a new destroyer as a follow-on or substitute for the Flight III DDG-51 that can support an SSL with a power level of 200 kW or 300 kW or more, and/or a megawatt-class FEL;[62]
- modify the designs of amphibious assault ships to be procured in coming years, so that they can support SSLs with power levels of 200 kW or 300 kW or more, and/or megawatt-class FELs; and
- modify the design of the Navy's new Ford (CVN-78) class aircraft carriers, if necessary, so that they can support SSLs with power levels of 200 kW or 300 kW or more, and/or megawatt-class FELs.[63]

An April 29, 2013, press report states:

> Now that the U.S. Navy is pushing even harder to equip its vessels with lasers, the service is focusing on reliable, high-voltage shipboard power to feed those weapons. Indeed, Navy officials say, meeting that need is becoming a matter of national security.
>
> "The work being done in this area is vital," said Thomas Killion, who heads the Office of Naval Research's (ONR's) Office of Transition, during this month's Electric Ship Technologies Symposium outside Washington. "As the upcoming deployment of a shipboard laser weapon reminds us, we need power generation and power management systems with greater-than-ever capabilities, but from devices that are smaller than ever."
>
> Now, navy scientists are looking for ways to better power those shipboard weapons. ONRsupported efforts are focused on cutting-edge technologies that include silicon carbide (SiC)- based transistors, transformers and power converters. "SiC is important because it improves power quality and reduces size and weight of components by as much as 90 percent," says Sharon Beerman-Curtin, ONR's power and energy science and technology lead. "This is a critical technology enabler for future Navy combatant ships that require massive amounts of highly controlled electricity to power advanced sensors, propulsion and weapons such as lasers and the electromagnetic railgun."[64]

OPTIONS FOR CONGRESS

Options for Congress regarding potential shipboard lasers include, among other things, the following:

- approve, reject, or modify the Navy's funding requests for development of potential shipboard lasers;
- request additional information from the Navy and DOD about potential shipboard lasers, perhaps by holding one or more hearings on the issue, or by requiring the Navy to submit one or more reports to Congress on the topic;
- review and comment on any roadmap for shipboard lasers that the Navy adopts;
- encourage or direct the Navy to adopt a program of record for procuring a production version of a shipboard laser;

- in the absence of a Navy program of record, direct the Navy to develop and install lasers with certain capabilities on a certain number of Navy surface ships by a certain date;[65]
- encourage or direct the Navy to design the Flight III version of the DDG-51 destroyer so that it can support an SSL with a power level of 200 kW or 300 kW or more;
- encourage or direct the Navy to design and procure a new destroyer as a follow-on or substitute for the Flight III DDG-51 that can support an SSL with a power level of 200 kW or 300 kW or more, and/or a megawatt-class FEL;
- encourage or direct the Navy to modify the designs of amphibious assault ships to be procured in coming years, so that they can support SSLs with power levels of 200 kW or 300 kW or more, and/or megawatt-class FELs; and
- encourage or direct the Navy to modify the design of the Navy's new Ford (CVN-78) class aircraft carriers, if necessary, so that they can support SSLs with power levels of 200 kW or 300 kW or more, and/or megawatt-class FELs.

LEGISLATIVE ACTIVITY FOR FY2016

FY2016 Funding Request

The Navy's proposed FY2016 budget requests $26.9 million for research and development work on directed energy technologies, including the SSL technologies, in the directed energy portion of Program Element (PE) 0602114N, Power Projection Applied Research, line item 4 in the Navy's FY0216 research and development account.[66] The figures for FY2014 and FY2015 for this PE are $40.3 million and $40.5 million, respectively. The Navy states that the "FY 2015 to FY 2016 decrease in funding is due to completion of the Solid State Laser - QRC [Quick Reaction Capability] program as well as a continued realignment of FEL activities."[67]

Additional funding for the Solid State Laser Technology Maturation Program (SSL-TM) forms part of the FY2016 funding request for the precision strike technology portion of PE 0603114N, Power Projection Advanced Technology, line item 15 in the Navy's FY2016 research and development account.[68]

The Navy's proposed FY2016 budget also requests $9.5 million for SSLs in Project 9823 (Lasers for Navy Applications) within PE 0603925N, Directed Energy and Electric Weapon System, line item 73 in the Navy's FY2016 research and development account.[69]

FY2016 National Defense Authorization Act (H.R. 1735/S. 1376)

House

The House Armed Services Committee, in its report (H.Rept. 114-102 of May 5, 2015) on H.R. 1735, recommends approving the Navy's FY2016 funding requests for PEs 0602114N, 0603114N, and 0603925N (page 461, lines 004 and 015, and page 463, line 063).

H.Rept. 114-102 states:

> Naval electric weapons systems fielding plan
>
> The committee is aware that the Navy has been pursuing development and operational demonstration of a number of electric weapons systems, including both directed energy systems and electromagnetic railguns. This class of electric weapons has the potential to provide revolutionary new capabilities for Navy platforms, including increased range, increased safety, and deeper magazines than conventional weapons. The committee believes that such systems will be important in the future to counter cost-imposing strategies in an anti-access environment where swarms of low-cost weapons could be used to overwhelm higher-cost, limited numbers of defensive weapons. However, as the Navy continues to pursue increasing power and decreasing size for such weapons, the committee believes that the Navy should also be considering how to field and integrate such systems into future naval platforms in order to facilitate successful transition from the laboratory to the fleet.
>
> Therefore, the committee directs the Secretary of the Navy to develop a plan for fielding electric weapon systems within the Department of the Navy for both the current and future fleet, and to provide a briefing on the results of this plan to the House Committee on Armed Services by March 1, 2016. As part of this plan, the Secretary of the Navy shall detail proposals for the allocation of the requisite power and space for the fielding of electric weapons systems, such as the Laser Weapons System, electromagnetic railgun, or other similar systems currently in development for the current and future fleet (Page 30).

Senate

The Senate Armed Services Committee, in its report (S.Rept. 114-49 of May 19, 2015) on S. 1376, recommends approving the Navy's FY2016 funding requests for PEs 0602114N, 0603114N, and 0603925N (page 403, line 4, page 404, line 15, and page 406, line 63).

Section 212 of S. 1376 as reported by the committee states (see in particular the parts in **bold**):

> SEC. 212. Department of Defense technology offset program to build and maintain the military technological superiority of the United States.
>
> (a) Program established.—
>
> (1) IN GENERAL.—The Secretary of Defense shall establish a technology offset program to build and maintain the military technological superiority of the United States by—
>
> (A) accelerating the fielding of offset technologies that would help counter technological advantages of potential adversaries of the United States, including directed energy, low-cost, high-speed munitions, autonomous systems, undersea warfare, cyber technology, and intelligence data analytics, developed using Department of Defense research funding and accelerating the commercialization of such technologies; and
>
> (B) developing and implementing new policies and acquisition and business practices.
>
> (2) GUIDELINES.—Not later than one year after the date of the enactment of this Act, the Secretary shall issue guidelines for the operation of the program, including—
>
> (A) criteria for an application for funding by a military department, defense agency, or a combatant command;
>
> (B) the purposes for which such a department, agency, or command may apply for funds and appropriate requirements for technology development or commercialization to be supported using program funds;
>
> (C) the priorities, if any, to be provided to field or commercialize offset technologies developed by certain types of Department research funding; and
>
> (D) criteria for evaluation of an application for funding or changes to policies or acquisition and business practices by a department, agency, or command for purposes of the program.
>
> **(b) Development of directed energy strategy.—**
>
> **(1) IN GENERAL.—Not later than one year after the date of the enactment of this Act, the Secretary, in consultation with such officials and third-party experts as the Secretary considers appropriate, shall develop a directed energy strategy to ensure that the United States directed energy technologies are being developed and deployed at an accelerated pace.**

(2) COMPONENTS OF STRATEGY.—The strategy required by paragraph (1) shall include the following:

(A) **A technology roadmap for directed energy that can be used to manage and assess investments and policies of the Department in this high priority technology area.**

(B) **Proposals for legislative and administrative action to improve the ability of the Department to develop and deploy technologies and capabilities consistent with the directed energy strategy.**

(C) **An approach to program management that is designed to accelerate operational prototyping of directed energy technologies and develop cost-effective, real-world military applications for such technologies.**

(3) BIENNIAL REVISIONS.—Not less frequently than once every 2 years, the Secretary shall revise the strategy required by paragraph (1).

(4) SUBMITTAL TO CONGRESS.—(A) Not later than 90 days after the date on which the Secretary completes the development of the strategy required by paragraph (1) and not later than 90 days after the date on which the Secretary completes a revision to such strategy under paragraph (3), the Secretary shall submit to the Committee on Armed Services of the Senate and the Committee on Armed Services of the House of Representatives a copy of such strategy.

(B) The strategy submitted under subparagraph (A) shall be submitted in unclassified form, but may include a classified annex.

(c) Applications for funding.—

(1) IN GENERAL.—Under the program, the Secretary shall, not less frequently than annually, solicit from the heads of the military departments, the defense agencies, and the combatant commands applications for funding to be used to enter into contracts, cooperative agreements, or other transaction agreements entered into pursuant to section 845 of the National Defense Authorization Act for Fiscal Year 1994 (Public Law 103–160; 10 U.S.C. 2371 note) with appropriate entities for the fielding or commercialization of technologies.

(2) TREATMENT PURSUANT TO CERTAIN CONGRESSIONAL RULES.—Nothing in this section shall be interpreted to require any official of the Department of Defense to provide funding under this section to any earmark as defined pursuant to House Rule XXI, clause 9, or any congressionally directed spending item as defined pursuant to Senate Rule XLIV, paragraph 5.

(d) Funding.—

(1) **IN GENERAL.—Subject to the availability of appropriations for such purpose, of the amounts authorized to be appropriated for research, development, test, and evaluation, Defense-wide for fiscal**

year 2016, not more than $400,000,000 may be used for any such fiscal year for the program established under subsection (a).

(2) **AMOUNT FOR DIRECTED ENERGY.—Of this amount, not more than $200,000,000 may be used for activities in the field of directed energy.**

(e) Transfer authority.—

(1) IN GENERAL.—The Secretary may transfer funds available for the program to the research, development, test, and evaluation accounts of a military department, defense agency, or a combatant command pursuant to an application, or any part of an application, that the Secretary determines would support the purposes of the program.

(2) SUPPLEMENT NOT SUPPLANT.—The transfer authority provided in this subsection is in addition to any other transfer authority available to the Department of Defense.

(f) Termination.—

(1) IN GENERAL.—The authority to carry out a program under this section shall terminate on September 30, 2020.

(2) TRANSFER AFTER TERMINATION.—Any amounts made available for the program that remain available for obligation on the date the program terminates may be transferred under subsection (e) during the 180-day period beginning on the date of the termination of the program.

Regarding Section 212, S.Rept. 114-49 states (see in particular the parts in **bold**):

Department of Defense technology offset program to build and maintain the military technological superiority of the United States (sec. 212)

The committee notes with concern that the United States has not faced a more diverse and complex array of crises since the end of World War II, and that taken together, they constitute the greatest challenge in a generation to the integrity of the liberal world order, which has consistently been underwritten by U.S. military technological superiority. At the same time, the committee is alarmed by the apparent erosion in recent years of this technological advantage, which is in danger of disappearing altogether. To prevent such a scenario and to maintain the country's global military technological edge, the committee recommends a provision that would establish a new $400.0 million initiative.

In doing so, the committee notes that the Defense Department is facing an emerging innovation gap. Commercial research and development in the United States now represents 80 percent of the national total, and the top four U.S. defense contractors combined spend only one-quarter of what the single biggest internet company does on research and development. Furthermore, global research and development is now more than twice that of the United States. The committee also notes that defense innovation is moving too slowly— in cycles that can last up to 18 years, whereas commercial innovation can be measured in cycles of 18 months or less.

The committee understands that accessing sources of innovation beyond the Defense Department is critical for national security, particularly in the areas of directed energy, low-cost high-speed munitions, cyber capabilities, autonomous systems, undersea warfare, and intelligence data analytics. However, there are currently too many barriers that limit cooperation with U.S. allies and global commercial firms, posing a threat to the country's future military technological dominance.

For the past several years, U.S. adversaries have been rapidly improving their own military capabilities to counter our unique advantages. Structural trends, such as the diffusion of certain advanced military technologies, pose new operational challenges to U.S. armed forces. As a result, the dominance of the United States military can no longer be taken for granted. Consequently, the Department of Defense must remain focused on the myriad potential threats of the future and thus maintain technological superiority against potential adversaries.

The committee notes that since 1960, the department has invested more than $6.0 billion in directed energy science and technology initiatives. The committee is concerned that, despite this significant investment, the department's directed energy initiatives are not resourced at levels necessary to transition them to full-scale acquisition programs. The committee is encouraged by the Navy's demonstration a 100–150 kilowatt prototype laser and by the Air Force's demonstration of high-powered electromagnetic weapons capabilities. However, the committee is concerned about the future of directed energy technologies as a whole. The committee notes that there is no inter-service entity dedicated to advancing promising directed energy platforms beyond the development point towards acquisition.

The committee is encouraged that the department established a department-wide Defense Innovation Initiative in November 2014 to pursue innovative ways to sustain and advance our military superiority and to improve business operations throughout the department. However, the committee is concerned by the possibility that this initiative is not being implemented in an appropriate and expeditious manner.

In response to these factors, the committee recommends a provision that would establish an initiative within the Department of Defense to maintain and enhance the military technological superiority of the United States. The provision would establish a program to accelerate the fielding of offset technologies, including,

but not limited to, directed energy, low-cost high-speed munitions, autonomous systems, undersea warfare, cyber technology, and intelligence data analytics, developed by the department and to accelerate the commercialization of such technologies. As part of this program, the committee expects that the Secretary of Defense would also establish updated policies and new acquisition and management practices that would speed the delivery of offset technologies into operational use.

The provision would authorize $400.0 million for fiscal year 2016 for the initiative, of which $200.0 million would be authorized specifically for directed energy technology. Accordingly, the provision would mandate the Secretary to develop a directed energy strategy to ensure that appropriate technologies are developed and deployed at an accelerated pace, and update it every 2 years. The committee expects that this strategy would include a recommendation on rationalizing the roles and authorities of the Joint Technology Office for High Energy Lasers. The provision would further direct the Secretary to submit this strategy to the Senate Armed Services Committee and the House Armed Services Committee no later than 90 days after completing the strategy, and biennially thereafter.

To speed up the development of these vitally needed national security capabilities, the committee directs that the Secretary of Defense shall consider all appropriate flexible acquisition authorities granted in law and in this Act. These should include the management structure and streamlined procedures for rapid prototyping outlined in section 803 of this Act on the middle tier of acquisition for rapid prototyping and rapid fielding, and the procedures and authorities to be considered under section 805 of this Act on use of alternative acquisition paths to acquire critical national security capabilities to include other transactions, rapid acquisition, and commercial item authorities.

The committee expects that the Secretary of Defense would keep the Senate Committee on Armed Services and the House Committee on Armed Services regularly updated on progress of activities under this technology offsets initiative (Pages 44-46).

APPENDIX A. LASER POWER LEVELS REQUIRED TO COUNTER TARGETS

Table A-1 shows two Navy perspectives, a Defense Science Board (DSB) task force perspective, and two industry perspectives on approximate laser power levels needed to affect various categories of targets.

Table A-1. Approximate Laser Power Levels Needed to Affect Certain Targets

Multiple perspectives that may reflect varying assumptions about BQ and other factors

Source	Beam power measured in kilowatts (kW) or megawatts (MW)				
	~ 10 kW	Tens of kW	~ 100 kW	Hundreds of kW	MW
One Navy briefing (2010)	UAVs				
		Small boats			
				Missiles (starting at 500 kW)	
Another Navy briefing (2010)		Short-range operations against UAVs, RAM, MANPADS (50 kW- 100kW; low BQ)		Extended-range operations against UAVs, RAM, MANPADS, ASCMs flying a crossing path (> 100 kW, BQ of ~ 2)	Operations against supersonic, highly maneuverable ASCMs, transonic air-to-surface missiles, and ballistic missiles (> 1 MW)
Industry briefing (2010)		UAVs and small boats (50 kW)	RAM (100 + kW), subsonic ASCMs (300 kW), manned aircraft (500 kW)		Supersonic ASCMs and ballistic missiles
Defense Science Board (DSB) report (2007)		Surface threats at 1-2 km		Ground-based air and missile defense, and countering rockets, artillery, and mortars, at 5-10 km[a]	"Battle group defense" at 5-20 km (1-3 MW)
Northrop Grumman research paper (2005)	Soft UAVs at short range	Aircraft and cruise missiles at short range	Soft UAVs at long range	Aircraft and cruise missiles at long range, and artillery rockets (lower hundreds of kW) Artillery shells and terminal defense against very short range ballistic missiles (higher 100s of kW)	

Source: One Navy briefing: Briefing slide entitled "HEL [High-Energy Laser] Missions," in briefing entitled "Directed Energy Warfare Office (DEWO) Overview," July 23, 2010. Another Navy briefing: Briefing slide entitled "Surface Navy Laser Vision," in briefing entitled "Navy Directed Energy Efforts – Ship Based Laser Weapon System," July 23, 2010.

Industry briefing: Briefing to CRS by an industry firm in the summer of 2010; data shown in table used here with the firm's permission. DSB report: *[Report of] Defense Science Board Task Force on Directed Energy Weapons*, December 2007, Table 2 (page 12). Northrop Grumman research paper: Richard J. Dunn, III, Operational Implications of Laser Weapons, Northrop Grumman (Analysis Center Papers), September 2005 (available online at http://www.northropgrumman.com/analysis-center/paper/assets/ Operational_Implications_of_La.pdf), visual inspection of Figure 1 (page 7).

Notes: kW is kilowatts; MW is megawatts; km is kilometer; RAM is rockets, artillery, mortars; MANPADS is man-portable air defense system (i.e., shoulder-fired surface-to-air missiles).

[a] Note that this statement refers to ground-based operations. It is not clear how this statement might change for shipboard operations, where atmospheric absorption due to water vapor can be an increased concern.

As can be seen in the table, these perspectives differ somewhat regarding the power levels needed to counter certain targets, perhaps because of differing assumptions about beam quality (BQ) and other factors.

APPENDIX B. NAVY ORGANIZATIONS INVOLVED IN DEVELOPING LASERS

Principal Navy organizations involved in developing lasers for potential shipboard use include

- the Office of Naval Research (ONR);
- the Naval Research Laboratory (NRL);
- the Directed Energy and Electric Weapon Systems (DE&EWS) Program Office (PMS-405);[70]
- the Naval Surface Warfare Center (NSWC) Dahlgren Division (NSWCDD), located at Dahlgren, VA; and
- the Directed Energy Warfare Office (DEWO), which the Navy established in August 2009 to serve as an NSWCDD center of excellence.

Additional Navy organizations involved in developing lasers for potential shipboard use include the CIWS program office (PEO IWS 3B, meaning Program Executive Officer, Integrated Warfare Systems, office code 3B); NSWC Crane Division at Crane, IN; NSWC Port Hueneme at Port Hueneme, CA; the Naval Air Weapon Stations at China Lake and Point Mugu, CA, as well as the Naval Air Station Patuxent River, MD, all of which are part of the Naval Air Systems Command (NAVIAR); and the Space and Naval Warfare Systems (SPARWAR) Center Pacific, located at San Diego.

Additional DOD organizations outside the Navy are also involved in developing lasers for potential shipboard use.

APPENDIX C. ADDITIONAL INFORMATION ON LASER WEAPON SYSTEM (LAWS)

A fiber SSL first uses high power semiconductor laser diodes to convert electricity into light. The light then passes through one or more glass optic fibers that contain a small amount of a deliberately introduced impurity, or "dopant" material, usually ytterbium (Yb). The interaction of the light with the dopant both changes the light's wavelength (color) and concentrates the light into a narrow laser beam that travels down the fiber until it exits the other end. Special optics combine the output of multiple fibers into one powerful beam. The fibers are referred to as the gain medium, and the laser is called a solid state laser because the gain medium is a solid rather than a liquid (such as in dye lasers) or a gas (as in gas lasers). Over the last decade, dramatic improvements in diodes and fiber materials have enabled a roughly 100-fold increase in the maximum power of an individual fiber SSL, from about 100 watts to about 10 kW.

The Navy's approach to developing LaWS was to maximize reliance on existing technology and components so as to minimize development and procurement costs. The LaWS prototype incoherently combines light beams from six fiber SSLs—commercial, off-the-shelf (COTS) welding lasers—each with a power of 5.5 kW, to create a laser with a total power of 33 kW[71] and a BQ of 17. The light from the six lasers is said to be incoherently combined because the individual beams are not merged into a true single beam (i.e., the individual beams are not brought in phase with each other). Although the beams are quite close to one another, they remain separate and out of phase with each other, and are steered and focused by the beam director so that they converge into a single spot when they reach the intended target. Coherently combining the six beams into a true single beam (i.e., one in which the six beams are "phase locked") would require a system with more-complex internal optics and electronic control systems.

LaWS, like many other fiber SSLs, emits light with a wavelength of 1.064 microns, which is close to, but not exactly at, an atmospheric transmission "sweet spot" at 1.045 microns.

LaWS is about 25% efficient, meaning that about 400 kW of ship's power would be needed to operate a future version of LaWS producing 100 kW of laser light. The remaining 300 kW of electrical energy would be converted into waste thermal energy (heat) that needs to be removed from the system using the ship's cooling capacity.

The conceptual breakthrough underpinning LaWS was made by scientists at the Pennsylvania State Electronic-Optic Center in 2004 and 2005 during some simple experiments, and by scientists at the Naval Research Laboratory (NRL) in 2006, in detailed analysis and subsequent experiments. Both groups showed that coherently combining light beams was not necessary to create a militarily useful laser from commercial fiber SSLs—that this could be done through the technically simpler approach of incoherently combining their beams.

DEWO is the lead system integrator (LSI) and technical direction agent for LaWS. Raytheon, the maker of CIWS, is the prime support contractor for the CIWS integration effort.[72]

A June 1, 2011, Navy information paper states:

> 1. The following efforts (funded under Fiscal Year [FY] 2010 Congressional Add) are underway to support the conduct of Trident Warrior (TW) 11 in the June 2011 timeframe:
>
> - Predictive Avoidance – continuing engineering, analysis, software development, and integration of a Predictive Avoidance Safety System (PASS) into the Prototype Laser Weapon System (LaWS)
> - Stabilization – continuing engineering, analysis, software development, and integration of Fast Steering Mirrors (FSM) as part of the Beam Control/Tracking subsystem of LaWS
> - LaWS KINETO Tracking Mount (KTM) Enclosure – material procurement and enclosure fabrication that will fit within the space constraints of the mechanized landing craft (LCM-8) as a test platform
> - TW 11 test planning and documentation development.
>
> Trident Warrior info can be found at: http://www.public.navy.mil/usff/tridentwarrior/Pages/default.aspx
>
> 2. The following efforts were accomplished or are underway in support of the PEO IWS Laser Close In Weapon System (CIWS) Draft Weapon Specification development effort:
>
> - Provided: Threat Vulnerability information for both in band and out of band laser engagements; LaWS test results from White Sands Missile Range and St. Nicholas Island; draft Design Reference Missions (DRMs); draft generic Concept of Operations (CONOPs); system level requirements for multi beam aperture system; draft space, weight, air, power requirements; Laser Trade Study Briefings.
> - In Process: Attending System Engineering Working Group (SEWG) meetings in support of Draft Weapon Specification development; providing technical reviews of Draft Weapon Specification developments.

3. The current Technology Readiness Level (TRL) of the Prototype LaWS is approaching 6, based on a system prototype demonstration in a relevant (maritime) environment.[73]

Figure C-1 shows a picture of the LaWS prototype; **Figure C-2** shows a rendering of LaWS when installed as an addition to a CIWS mount. In **Figure C-2**, the red-colored tube hanging off the left side of the CIWS mount is the LaWS beam director, and the white device bolted to the right side of the CIWS radome is another LaWS component.

Source: Photograph provided by Navy Office of Legislative Affairs, November 3, 2010.

Figure C-1. Photograph of LaWS Prototype.

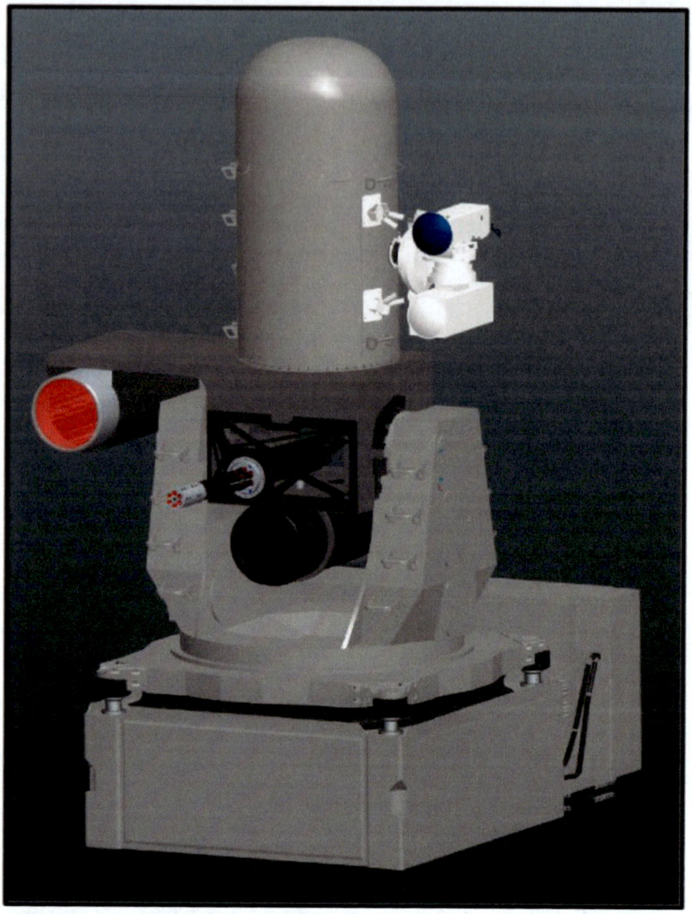

Source: Rendering provided by Navy Office of Legislative Affairs, November 3, 2010. In this rendering, the red-colored tube hanging off the left side of the CIWS mount is the LaWS beam director, and the white device bolted to the right side of the CIWS radome is another LaWS component.

Figure C-2. Rendering of LaWS Integrated on CIWS Mount.

APPENDIX D. ADDITIONAL INFORMATION ON TACTICAL LASER SYSTEM (TLS)

A June 10, 2011, Navy information paper states:

The MK 38 TLS concept is based on a commercial off-the-shelf (COTS) Solid State Laser (SSL) with a simple Beam Director (BD) integrated with the MK 38 Mod 2 Machine Gun System (MGS). Other high energy SSL typically combine several individual beams in order to achieve a higher power output. The TLS is a single phase laser, meaning it does not utilize a combination of several lasers. This does reduce the total power output of the system, but allows for a far greater Beam Quality (BQ). The current BQ is 2.1, but modifications are being made to improve this to 1.5. Beam Quality, along with power output, is a key parameter to determining a laser's effectiveness against targets. With the current BQ and power output, the TLS should be capable of defeating some small boat targets at ranges of up to 2 km, given optimal weather and sea conditions. A future demonstration of the laser system's effectiveness is currently planned in March 2012.

The BD is a simple design with relatively few moving parts. Independent drives enable the TLS to make azimuth corrections faster and point beyond the elevation limits of the MK 38 Mod 2 MGS. The current integration work for the TLS is to have the MK 38 Mod 2 MGS Electro-Optical Sight (EOS) hand track over to the TLS. Track handoff from the EOS to the TLS will be tested in an event scheduled for 29 June 2011 at Eglin Air Force Base.

The TLS is about 30% efficient, meaning 34 kW of power is needed to operate the 10 kW laser. The remaining 24 kW are converted into thermal energy that must be removed from the system. Currently, the TLS will utilize its own power distribution and cooling systems. The power requirement from a ship would be approximately 75 kW, 440 VAC 60 Hz 3 Phase power to run the laser, power management, and currently installed/designed thermal management systems. Additional engineering development would be required for actual shipboard use.

Technical risks identified for the TLS demonstration [include] MGS integration, laser Beam Quality and Beam Director tracking. Accurate target range data is critical to the effectiveness of the TLS. The BD does not include a Laser Range Finder (LRF), and the MK 38 EOS is expected to provide this data. The interface of the EOS and TLS will be tested in June at Eglin as mentioned above. A failure to improve BQ or demonstrate stable tracking for the BD, will impact system effectiveness resulting in reduced range and higher laser dwell times.

IPG is the COTS laser manufacturer. Boeing is the BD designer and Laser Weapons Module lead. The MK 38 system integrator is BAE Systems.[74]

Figure D-1 shows a rendering of TLS when installed as an addition to the Mk 38 machine gun system.

Source: BAE news release dated April 7, 2011, entitled "BAE Systems Selected to Demonstrate Tactical Laser System for the U.S. Navy," accessed online July 5, 2011, at http://www.baesystems.com/Newsroom/ NewsReleases/autoGen_ 1113718157.html.

Figure D-1. Rendering of TLS Integrated on Mk 38 Machine Gun Mount.

APPENDIX E. ADDITIONAL INFORMATION ON MARITIME LASER DEMONSTRATION (MLD)

Slab SSLs are similar to fiber SSLs, except that the synthetic crystalline material used as the gain medium is formed into plate-like slabs rather than flexible fibers. Slab SSLs are being developed not just by the Navy, but by other U.S. military services, permitting the Navy to leverage development work funded by other parts of DOD.

MLD coherently combines beams from multiple slab SSLs, each with a power of 15 kW, to create a higher-power beam with a good BQ. Each 15 kW laser is housed in a Line Replaceable Unit (LRU) measuring about 1 foot by 2 feet by 3.5 feet. MLD might be installed on its own mount rather than as an addition to a ship's existing CIWS mount.

MLD, like LaWS, emits light with a wavelength of 1.064 microns, which is close to, but not exactly at, an atmospheric transmission "sweet spot" at 1.045 microns.

Slab SSLs are currently about 20% to 25% efficient, meaning that about 400 kW to 500 kW of a ship's power would be needed to operate a system producing 100 kW of laser light. The remaining 300 kW to 400 kW of electrical energy would be converted into waste thermal energy that needs to be removed from the system using the ship's cooling capacity. Future slab SSLs might have efficiencies of about 30%.

In March 2009, Northrop demonstrated a version of MLD that coherently combined seven slab SSLs, each with a power of about 15 kW, to create a beam with a power of about 105 kW and a BQ of less than 3.[75]

Scaling up a slab laser to a total power of 300 kW and a BQ of 2 is not considered to require any technological breakthroughs. A slab laser with a total power of 300 kW might require a below-deck space measuring roughly 4.5 feet by 8 feet by 12 feet. Supporters of slab SSLs such as MLD believe they could eventually be scaled up further, to perhaps 600 kW. Slab SSLs are not generally viewed as easily scalable to megawatt power levels.

MLD is a commercially integrated weapon system with Northrop and L3-Brashears as the principal contractors. The government test team includes NSWC Dahlgren (VA), NSWC Port Hueneme (CA), and NAWC China Lake (CA). Although Northrop is the primary contractor for MLD, several other firms, such as Raytheon and Textron, are involved in efforts to develop slab SSLs for potential use by U.S. military services.

An April 8, 2011, ONR news release stated:

> Marking a milestone for the Navy, the Office of Naval Research and its industry partner on April 6 successfully tested a solid-state, high-energy laser (HEL) from a surface ship, which disabled a small target vessel.
>
> The Navy and Northrop Grumman completed at-sea testing of the Maritime Laser Demonstrator (MLD), which validated the potential to provide advanced self-defense for surface ships and personnel by keeping small boat threats at a safe distance.
>
> "The success of this high-energy laser test is a credit to the collaboration, cooperation and teaming of naval labs at Dahlgren, China Lake, Port Hueneme and Point Mugu, Calif.," said Chief of Naval Research Rear Adm. Nevin Carr. "ONR coordinated each of their unique capabilities into one cohesive effort."
>
> The latest test occurred near San Nicholas Island, off the coast of Central California in the Pacific Ocean test range. The laser was mounted

onto the deck of the Navy's self-defense test ship, former USS Paul Foster (DD 964).

Carr also recognized the Office of the Secretary of Defense's High Energy Joint Technology Office and the Army's Joint High Powered Solid State Laser (JHPSSL) program for their work. MLD leverages the Army's JHPSSL effort.

"This is the first time a HEL, at these power levels, has been put on a Navy ship, powered from that ship and used to defeat a target at-range in a maritime environment," said Peter Morrison, program officer for ONR's MLD.

In just slightly more than two-and-a-half years, the MLD has gone from contract award to demonstrating a Navy ship defensive capability, he said.

"We are learning a ton from this program—how to integrate and work with directed energy weapons," Morrison said. "All test results are extremely valuable regardless of the outcome."

Additionally, the Navy accomplished several other benchmarks, including integrating MLD with a ship's radar and navigation system and firing an electric laser weapon from a moving platform at-sea in a humid environment. Other tests of solid state lasers for the Navy have been conducted from land-based positions.

Having access to a HEL weapon will one day provide warfighter with options when encountering a small-boat threat, Morrison said.

But while April's MLD test proves the ability to use a scalable laser to thwart small vessels at range, the technology will not replace traditional weapon systems, Carr added.

"From a science and technology point of view, the marriage of directed energy and kinetic energy weapon systems opens up a new level of deterrence into scalable options for the commander. This test provides an important data point as we move toward putting directed energy on warships. There is still much work to do to make sure it's done safely and efficiently," the admiral said.[76]

A June 1, 2011, Navy information paper states:

As part of [ONR's] Survivability and Self Defense focus area, ONR with NAVSEA Program Executive Office for Integrated Weapons Systems (PEO IWS), the NAVSEA Directed Energy Program Office (PMS-405), the DoD High Energy Laser Joint Technology Office (HEL JTO) and the US Army Space and Missile Development Command (USA/SMDC), contracted with Northrop Grumman to design, develop, integrate, install and test the Maritime Laser Demonstration (MLD) from 2009 until early 2011.

The MLD program's main objective was to demonstrate a ship based laser "proof-ofconcept" weapons system to defend against small boat attacks, using commercially available laser and beam director components. The demonstrator showed the system design could be installed and function on existing Navy DDG, CG, LSD, LPD, LHA, LHD, and/or FFG ships; using the ship's power and fire control capabilities, and use advanced solid state laser slab directed energy technologies similar to those used in industrial applications. The successful testing and temporary integration of the MLD on the USS Paul Foster (US Navy Spruance Class test ship) and the acquired experience promotes confidence in the ability to subsequently develop a notional Naval Maritime Laser based Weapon System (NMLWS). The MLD Program marked a significant new naval capability to deter and inhibit an attack by small fast attack boats in a maritime environment.

After testing, the MLD system was removed from the USS Paul Foster and returned to Northrop Grumman facilities in El Segundo, California. The MLD system, as tested, employed a 15 Kilowatt 1.065 micron wavelength laser developed in the OSD HEL JTO Joint High Power Solid State Laser (JHPSSL) program, and on loan from the USA/SMDC. The modified JHPSSL module's output was directed to the target boat and laser fluence on the target was controlled by a motion stabilized beam director. Initial tracking of high speed, remotely operated and maneuverable small boat surface targets was provided by the ship's complement of existing radars, and then passively and actively tracked by the beam director cameras through varying environmental conditions up to World Meteorological Organization (WMO) sea states of three (3). Active engagement of the target was controlled by test, safety and fire controllers on the USS Paul Foster, located in the ship's command center. Significant data collection and photo coverage was gained during testing. In early April of 2011, the Maritime Laser Demonstration program showed significant capabilities for defeating small boats through the defeat of structural elements of the small boat. Additionally, engines on the remotely operated small boat target were later set ablaze by the laser at distances of over one mile. The MLD program marks the first time a laser weapon has been test fired from a US Navy ship, and successfully showed the potential power of a laser weapon system in the maritime environment.

The unclassified and publically released video of the testing of the MLD system may be viewed at YouTube^tm at the URL: http://www.you tube.com/watch?v=awsQs4ct0c4.[77]

Figure E-1 shows the MLD on a trailer; Figure E-2 shows a schematic of the system; Figure E3 shows a rendering of the beam director for the MLD in a notional shipboard installation.

Source: Photograph provided by Navy, November 29, 2010.

Figure E-1. Photograph of MLD on Trailer.

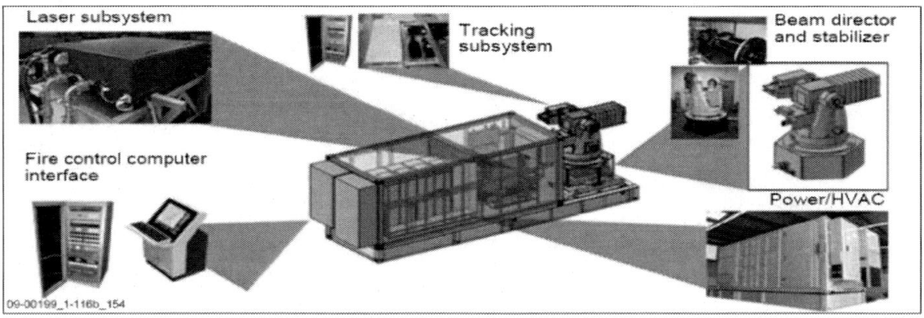

Source: Illustration provided by Navy, November 11, 2010.

Figure E-2. Schematic of MLD.

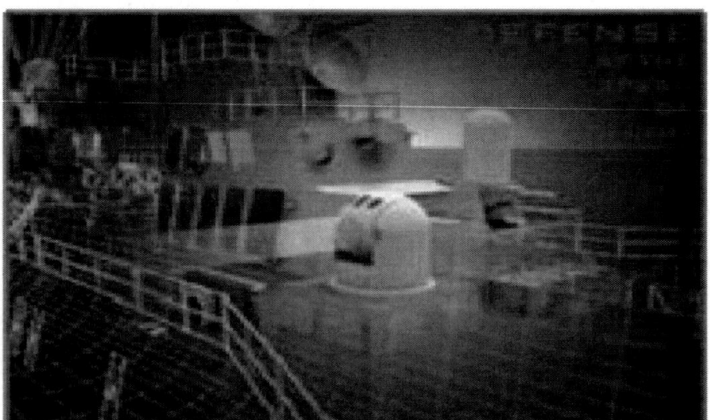

Source: Photograph provided by Northrop, October 21, 2010.

Figure E-3. Rendering of MLD in Notional Shipboard Installation.

APPENDIX F. ADDITIONAL INFORMATION ON FREE ELECTRON LASER (FEL)

An FEL uses an electron gun to generate a stream of electrons. The electrons are then sent into a linear particle accelerator to accelerate them to light speeds. The accelerated electrons are then sent into a device, known informally as a wiggler, that exposes the electrons to a transverse magnetic field, which causes the electrons to "wiggle" from side to side and release some of their energy in the form of light (photons). The photons are then bounced between mirrors and emitted as a coherent beam of laser light. To increase the efficiency of the system, some of the electrons are then cycled back to the front of the particle accelerator via an energy recovery loop.[78]

Unlike an SSL, which emits light with a fixed wavelength determined by the composition of its gain medium, an FEL's components can be adjusted to change the wavelength of light that it emits, so as to match various atmospheric transmission "sweet spots." The basic architecture of an FEL offers a clear potential for scaling up to power levels of one or more megawatts. A well-designed FEL can in theory be increased in power from 10 kW to 1 MW without an increase in system size, and without need for beam combiners. An FEL emits a beam with a BQ of 1 or close to 1.

Schematics of notional or developmental shipboard FELs today generally show them as devices with a length of roughly 100 feet. An FEL's ultimate shipboard space requirements will depend in part on how it is integrated into a ship's design, and whether the FEL uses room-temperature or superconducting particle-acceleration structures. Using superconducting acceleration structures can reduce the length of an FEL, and would require the use of cryogenic equipment to bring the superconducting structures down to the very low temperatures needed to make them superconducting. Operating an FEL would result in the production of X rays, requiring the system to be shielded to protect the ship's crew and other parts of the ship.

FELs that recycle electrons have an efficiency of about 10%, meaning that about 10 MW of ship's power would be needed to operate an FEL producing 1 MW of laser light. The remaining 9 MW of electrical energy is converted into waste thermal energy.

The FEL development effort is led by ONR. The effort also includes several other Navy organizations and institutions,[79] four Department of Energy (DOE) laboratories,[80] and several universities.[81] Contractors involved in FEL development have included Boeing (CA), Raytheon (MA), SAIC (VA),

Niowave (MI), and Advanced Energy Systems (NY). Boeing and Raytheon competed for the contract to design the 100 kW FEL. In September 2010, ONR announced that it had selected Boeing.[82] The award makes Boeing the Navy's current primary contractor for FEL development.

A January 20, 2011, news report states:

> Scientists at Los Alamos National Lab in Los Alamos, N.M., have achieved a remarkable breakthrough with the Office of Naval Research's (ONR) Free Electron Laser (FEL) program, setting the stage for a preliminary design review scheduled Jan. 20-21 in Virginia.
>
> Researchers demonstrated an injector capable of producing the electrons needed to generate megawatt-class laser beams for the Navy's next-generation weapon system Dec. 20, months ahead of schedule.
>
> "The injector performed as we predicted all along," said Dr. Dinh Nguyen, senior project leader for the FEL program at the lab. "But until now, we didn't have the evidence to support our models. We were so happy to see our design, fabrication and testing efforts finally come to fruition. We're currently working to measure the properties of the continuous electron beams, and hope to set a world record for the average current of electrons."
>
> Quentin Saulter, FEL program manager for ONR, said the implications of the FEL's progress are monumental.
>
> "This is a major leap forward for the program and for FEL technology throughout the Navy," said Saulter. "The fact that the team is nine months ahead of schedule provides us plenty of time to reach our goals by the end of 2011."[83]

A June 1, 2011, Navy information paper states:

> In September 2010, Boeing was selected as the lead systems integrator for the critical design phase of the FEL INP to design, develop, integrate and test a 100kw Free Electron Laser demonstration prototype that will be used to study scaling to megawatt level output powers. Boeing successfully completed the Preliminary Design Review in January 2011 and is working on the critical design of the 100kW demonstration prototype.
>
> The Navy's goal is to build a megawatt-class free electron laser that due to its flexibility in operating at multiple wavelengths has more capability than any other HEL weapon system to operate in any maritime environment in the world. Its all electric nature and multimission capability could reduce the cost and logistics burden for the Navy. Presently the FEL program is the only peer-reviewed electric laser megawatt class program in DoD.[84]

A March 21, 2012, press report stated that the FEL project was undergoing critical design review (CDR) that week.[85] A March 26, 2012, press report stated that "Boeing made good progress maturing the megawatt free electron laser, as shown during its critical design review.... " The report stated: "[Roger] McGinnis, [program executive for INPs at ONR's Naval Air Warfare and Weapons Department], said that the optics was likely the most challenging part but added that Boeing's optics system looked very good during the CDR."[86]

Figure F-1 shows part of an FEL facility at the Thomas Jefferson National Laboratory (Jefferson Lab) in Virginia. **Figure F-2** shows a simplified diagram of how an FEL works. **Figure F-3** shows a Jefferson Lab schematic of an FEL equipped with two "wigglers"—one for producing infrared (IR) laser light, and one for producing ultraviolet (UV) laser light. The FEL being developed by the Navy for shipboard use would likely produce only infrared light.

Source: Jefferson Lab news release of July 30, 2004, entitled "FEL Achieves 10 Kilowatts," accessed November 16, 2010 at http://www.jlab.org/news/releases /2004/0410kw.html. The news release says that the release is "As released by the Office of Naval Research with images and captions from Jefferson Lab." The caption to the photo in the news release states: "The Free-Electron Laser vault at Jefferson Lab showing the superconducting accelerator in the background and the magnetic wiggler in the foreground. The wiggler converts the electron beam power into laser light. Photo by Greg Adams, JLab."

Figure F-1. Photograph of an FEL Facility.

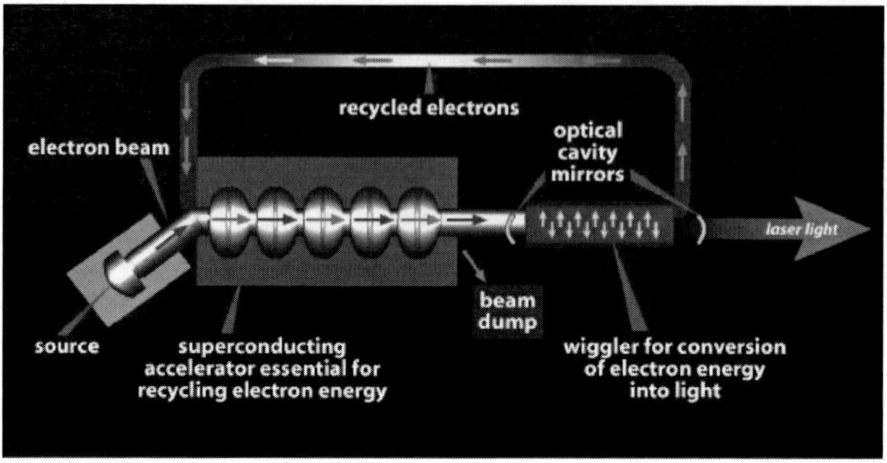

Source: Jefferson Lab web page providing an introduction to FELs, accessed November 16, 2010, at http://www.jlab.org/FEL/feldescrip.html.

Figure F-2. Simplified Diagram of How an FEL Works.

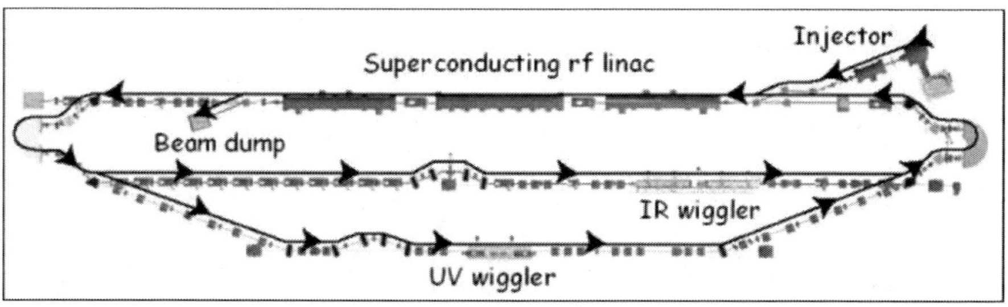

Source: Jefferson Lab web page describing its FEL, accessed November 16, 2010 at http://www.jlab.org/FEL/ felspecs.html. This FEL has two "wigglers"—one for producing infrared (IR) laser light, and one for producing ultraviolet (UV) laser light. The FEL being developed by the Navy for shipboard use would likely produce only infrared light. The arrows show the flow of electrons in the device, starting with the electron gun and injector in the upper-right corner. "Rf linac" means radio frequency linear accelerator.

Figure F-3. Schematic of an FEL; (Version with two "wigglers").

APPENDIX G. INNOVATIVE NAVAL PROTOTYPES (INPS)

The Office of Naval Researach (ONR) is developing the 100 kW FEL as an Innovative Naval Prototype (INP). ONR describes INPs as follows:

> [ONR's work on] Leap Ahead Innovations include Innovative Naval Prototypes (INPs) and Swampworks, and are technology investments that are potentially "game changing" or "disruptive" in nature. INPs achieve a level of technology suitable for transition in four to eight years. Innovative Naval Prototypes explore high 6.2 and 6.3 [research and development budget category] technologies that can dramatically change the way Naval forces fight. Programs in this category may be disruptive technologies that, for reasons of high risk or radical departure from established requirements and concepts of operation, are unlikely to survive without top leadership endorsement, and, unlike Future Naval Capabilities [another category of ONR's work], are initially too high risk for a firm transition commitment from the acquisition community. INPs should be identified based on a balanced combination of naval need and technology exploitation. Investments should be planned with the critical mass needed to achieve a level of technology maturity suitable for transition in four to eight years. Program Managers (PMs) are primarily selected from ONR, and in order to help facilitate the transition to the acquisition community, Deputy PMs are typically chosen from the Acquisition community. The CNR [Chief of Naval Research], in consultation with senior Navy and Marine Corps leadership, identifies candidate INPs that are then forwarded to Naval S&T [Science and Technology] Corporate Board (ASN-RDA, VCNO and the ACMC) [the Assistant Secretary of the Navy, Research, Development, and Acquisition, the Vice Chief of Naval Operations, and the Assistant Commandant of the Marine Corps] for approval / disapproval. Free Electron Laser is an innovative naval prototype. Swampworks efforts are smaller in scope than INPs and are intended to produce results in one to three years. This category is where we typically accept higher risk in an effort to produce higher payoff for the warfighters.[87]

APPENDIX H. DOD TECHNOLOGY READINESS LEVELS (TRLS)

DOD uses TRLs to characterize the developmental status of many weapon technologies. DOD defines its TRLs as follows:

- TRL 1: Basic principles observed and reported.
- TRL 2: Technology concept and/or application formulated.
- TRL 3: Analytical and experimental critical function and/or characteristic proof of concept.

- TRL 4: Component and/or breadboard validation in a laboratory environment.
- TRL 5: Component and/or breadboard validation in a relevant environment.
- TRL 6: System/subsystem model or prototype demonstration in a relevant environment.
- TRL 7: System prototype demonstration in an operational environment.
- TRL 8: Actual system completed and qualified through test and demonstration.
- TRL 9: Actual system proven through successful mission operations.[88]

APPENDIX I. PROTOCOL ON BLINDING LASERS

This appendix provides information on the international protocol on blinding lasers and its relationship to DOD laser programs, including the lasers discussed in this report.

Overview

The United States in 1995 ratified the 1980 Convention on Prohibitions or Restriction on the Use of Certain Conventional Weapons Which May be Deemed to be Excessively Injurious or to Have Indiscriminate Effects. An international review of the convention began in 1994 and concluded in May 1996 with the adoption of, among other things, a new Protocol IV on blinding laser weapons. The protocol prohibits the employment of lasers that are specifically designed to cause permanent blindness to the naked eye or to the eye with corrective eyesight devices.

The United States ratified Protocol IV on December 23, 2008, and it entered into force for the United States on July 21, 2009.[89] DOD views the protocol as fully consistent with DOD policy. DOD believes the lasers discussed in this report are consistent with DOD policy of prohibiting the use of lasers specifically designed to cause permanent blindness to the naked eye or to the eye with corrective eyesight devices.

Article-by-Article Discussion

Article 1 of the protocol prohibits the employment of "laser weapons specifically designed, as their sole combat function or as one of their combat functions, to cause permanent blindness to unenhanced vision, that is to the naked eye or to the eye with corrective eyesight devices." DOD states that:

> This prohibition is fully consistent with the policy of the Department of Defense, which is to prohibit the use of weapons so designed. Although the prospect of mass blinding was an impetus for the adoption of the Protocol, it was not the intent of the Conference to prohibit only mass blinding. Accordingly, under both the Blinding Laser Protocol and Department of Defense policy, laser weapons designed specifically to cause such permanent blindness may not be used against an individual enemy combatant.[90]

Article 2 of the protocol obligates parties to "take all feasible precautions to avoid the incidence of permanent blindness to unenhanced vision." DOD states that "This requirement is also fully consistent with the policy of the Department of Defense which is to reduce, through training and doctrine, inadvertent injuries from the use of lasers designed for other purposes, such as range-finding, target discrimination, and communications."[91]

Article 3 of the protocol states that "blinding as an incidental or collateral effect of the legitimate military employment of laser systems, including laser systems used against optical equipment, is not covered" by the Protocol. DOD states that this article "reflects a recognition of the inevitability of eye injury as the result of lawful battlefield laser use. Its use is an important measure in avoiding war crimes allegations where injury occurs from legitimate laser uses."[92]

DOD further states that

> As a matter of policy, the United States will refrain from the use of laser weapons prohibited by the Protocol. Therefore, while the Blinding Laser Weapons Protocol does not legally apply to all armed conflicts, it is U.S. policy to apply the Protocol to all such conflicts, however they may be characterized, and in peacetime.... The Protocol is fully consistent with U.S. military interests, Department of Defense policy and humanitarian concerns generally. Accordingly, the United States should ratify it at an early date.[93]

Excerpt from 2007 DSB Task Force Report

A 2007 report by a Defense Science Board (DSB) task force on directed energy weapons stated:

> The task force heard concerns over the legal and policy aspects of employing directed energy weapons. The concern is seen by some as inhibiting or deterring development of such weapons with [i.e., due to] reluctance to invest in capabilities that might not be useable in the battlespace due to legal or policy constraints. Much of this concern is the product of inadequate communications rather than any unusual legal or policy constraints.
>
> The Office of the Secretary of Defense and service component Judge Advocate General Offices have determined that directed energy weapons are, in and of themselves, legal according to all U.S. laws, [as well as] the [international] Laws of Armed Conflict, and are consistent with all current U.S. treaty and international obligations. Noting that directed energy weapons are legal does not imply that their use in a particular situation is legal. There are situations where the use of a directed energy weapon could be contrary to U.S. or international law. This consideration is the case with virtually any weapon.
>
> One such constraint is the use of a laser weapon to intentionally blind combatants. The States Parties to the 1980 Convention on Prohibitions or Restrictions on the use of Certain Conventional Weapons Which May Be Deemed to be Excessively Injurious or to have Indiscriminate Effects had a fourth protocol adopted in 1995, where the intent is to prohibit laser weapons that are specifically used to blind combatants systematically and intentionally. While the United States is not a signatory to this particular protocol, the DOD has issued a policy that prohibits the use of lasers specifically designed to cause permanent blindness of unenhanced vision.
>
> That same policy stated that "... laser systems are absolutely vital to our modern military. Among other things, they are currently used for detection, targeting, range-finding, communications, and target destruction. They provide a critical technological edge to U.S. forces and allow our forces to fight, win, and survive on an increasingly lethal battlefield. In addition, lasers provide significant humanitarian benefits. They allow weapon systems to be increasingly discriminate, thereby reducing collateral damage to civilian lives and property. The [DOD] recognizes that accidental or incidental eye injuries may occur on the battlefield as the result of the use of legitimate laser systems.[94] Therefore, we continue to strive, through training and doctrine, to minimize these injuries."
>
> A similarly supportive policy has been stated for other directed energy weapons. At the same time, when such weapons are new to the

battlespace, there will be a policy determination on their initial introduction to include an understanding by appropriate policy makers of the intended uses. Such determination needs to be informed by a thorough and credible understanding of the risk and benefits of employing such weapons. Beyond the process of approving first use, the expectation is that the Laws of Armed Conflict, rules of engagement, and combat commander direction will govern employment of directed energy weapons as is the case for kinetic weapons.[95]

APPENDIX J. ILLUMINATION OF OBJECTS IN SPACE

In briefings on potential shipboard lasers, Navy officials noted DOD Instruction (DODI) 3100.11 of March 31, 2000, which states in part:

> All DoD laser activities shall be conducted in a safe and responsible manner that protects space systems, their mission effectiveness, and humans in space, consistent with national security requirements, in accordance with [DoD Directive 3100.10, "Space Policy," July 9, 1999]. All such activities shall be coordinated with the Commander in Chief of U.S. Space Command (CINCSPACE) for predictive avoidance or deconfliction with U.S., friendly, and other space operations.[96]

The technical community in the Navy believes that this instruction effectively requires the military services to implement measures for ensuring that objects in space face low or no exposure to laser energy. The technical community believes that this in turn would require that shipboard lasers incorporate so-called predictive avoidance (PA) software and/or other features that would prevent them from firing in the direction of an object in space. The community believes that two policy changes would be required to permit Navy surface ships to use shipboard lasers with power levels high enough that they could cause unwanted collateral damage to satellites:

- The community believes that current safety criteria relating to satellites are overly restrictive and should be replaced with a new policy that includes what the Navy views as more realistic safety criteria.
- The community believes that certain data relating to sensitive satellites should be removed from the PA system so that the classification level of the PA system can be lowered.

APPENDIX K. SECTION 220 OF FY2000 DEFENSE AUTHORIZATION ACT (P.L. 106-398)

As mentioned earlier (see footnote footnote 65 in "Options for Congress"), the option of directing the Navy to develop and install lasers with certain capabilities on a certain number of Navy surface ships by a certain date could take the form of a provision broadly similar to Section 220 of the FY2001 defense authorization act (H.R. 4205/P.L. 106-398 of October 30, 2000), which set goals for the deployment of unmanned combat aircraft and unmanned combat vehicles. The text of Section 220 is as follows:

> SEC. 220. UNMANNED ADVANCED CAPABILITY COMBAT AIRCRAFT AND GROUND COMBAT VEHICLES.
>
> (a) GOAL- It shall be a goal of the Armed Forces to achieve the fielding of unmanned, remotely controlled technology such that—
>
> by 2010, one-third of the aircraft in the operational deep strike force aircraft fleet are unmanned; and
>
> by 2015, one-third of the operational ground combat vehicles are unmanned.
>
> (b) REPORT ON UNMANNED ADVANCED CAPABILITY COMBAT AIRCRAFT AND GROUND COMBAT VEHICLES- (1) Not later than January 31, 2001, the Secretary of Defense shall submit to the congressional defense committees a report on the programs to demonstrate unmanned advanced capability combat aircraft and ground combat vehicles undertaken jointly between the Director of the Defense Advanced Research Projects Agency and any of the following:
>
> (A) The Secretary of the Army.
>
> (B) The Secretary of the Navy.
>
> (C) The Secretary of the Air Force.
>
> (2) The report shall include, for each program referred to in paragraph (1), the following:
>
> (A) A schedule for the demonstration to be carried out under that program.
>
> (B) An identification of the funding required for fiscal year 2002 and for the future-years defense program to carry out that program and for the demonstration to be carried out under that program.
>
> (C) In the case of the program relating to the Army, the plan for modification of the existing memorandum of agreement with the Defense Advanced Research Projects Agency for demonstration and development of the Future Combat System to reflect an increase in unmanned, remotely controlled enabling technologies.
>
> (3) The report shall also include, for each Secretary referred to in paragraphs (1)(A), (1)(B), and (1)(C), a description and assessment of the

acquisition strategy for unmanned advanced capability combat aircraft and ground combat vehicles planned by that Secretary, which shall include a detailed estimate of all research and development, procurement, operation, support, ownership, and other costs required to carry out such strategy through the year 2030, and—

(A) in the case of the acquisition strategy relating to the Army, the transition from the planned acquisition strategy for the Future Combat System to an acquisition strategy capable of meeting the goal specified in subsection (a)(2);

(B) in the case of the acquisition strategy relating to the Navy—

(i) the plan to implement a program that examines the ongoing Air Force unmanned combat air vehicle program and identifies an approach to develop a Navy unmanned combat air vehicle program that has the goal of developing an aircraft that is suitable for aircraft carrier use and has maximum commonality with the aircraft under the Air Force program; and

(ii) an analysis of alternatives between the operational deep strike force aircraft fleet and that fleet together with an additional 10 to 20 unmanned advanced capability combat aircraft that are suitable for aircraft carrier use and capable of penetrating fully operational enemy air defense systems; and

(C) in the case of the acquisition strategy relating to the Air Force—

(i) the schedule for evaluation of demonstration results for the ongoing unmanned combat air vehicle program and the earliest possible transition of that program into engineering and manufacturing development and procurement; and

(ii) an analysis of alternatives between the currently planned deep strike force aircraft fleet and the operational deep strike force aircraft fleet that could be acquired by fiscal year 2010 to meet the goal specified in subsection (a)(1).

(c) FUNDS- Of the amount authorized to be appropriated for Defense-wide activities under section 201(4) for the Defense Advanced Research Projects Agency, $100,000,000 shall be available only to carry out the programs referred to in subsection (b)(1).

(d) DEFINITIONS- For purposes of this section:

(1) An aircraft or ground combat vehicle has ‘unmanned advanced capability’ if it is an autonomous, semi-autonomous, or remotely controlled system that can be deployed, re-tasked, recovered, and re-deployed.

(2) The term ‘currently planned deep strike force aircraft fleet’ means the early entry, deep strike aircraft fleet (composed of F-117 stealth aircraft and B-2 stealth aircraft) that is currently planned for fiscal year 2010.

(3) The term ‘operational deep strike force aircraft fleet’ means the currently planned deep strike force aircraft fleet, together with at least 30

unmanned advanced capability combat aircraft that are capable of penetrating fully operational enemy air defense systems.

(4) The term 'operational ground combat vehicles' means ground combat vehicles acquired through the Future Combat System acquisition program of the Army to equip the future objective force, as outlined in the vision statement of the Chief of Staff of the Army.

End Notes

[1] Lara Seligman, "Navy-built LaWS To Begin Demo This Summer, IOC Slated For FY-20-21," *Inside the Navy*, March 24, 2014. A program of record, or POR, is a term sometimes used by DOD officials that means, in general, a program in the Future Years Defense Plan (FYDP) that is intended to provide a new, improved, or continuing materiel, weapon, or information system or service capability in response to an approved need. The term is sometimes used to refer to a program in a service's budget for procuring and deploying an operational weapon system, as opposed to a research and development effort that might or might not eventually lead to procurement and deployment of an operational weapon system. If a research and development effort is converted into a program or record for procuring an operational weapon system, the program might then be conducted under the DOD's process for managing the acquisition of weapon systems, which is discussed further in CRS Report RL34026, *Defense Acquisitions: How DOD Acquires Weapon Systems and Recent Efforts to Reform the Process*, by Moshe Schwartz.

[2] In discussions of other types of defense systems, the terms short range and long range could have considerably different meanings. In discussions of the ranges of military airplanes or ballistic missiles, for example, the term short range might mean a range of hundreds of miles, while references to longer ranges could refer to ranges of thousands of miles.

[3] See, for example, Geoff Fein, "Navy Leveraging Commercial Lasers To Shoot Down UAVs," *Defense Daily*, May 11, 2010: 3-4.

[4] A "hard kill" involves destroying the attacking weapon in some manner. A "soft kill" involves confusing the weapon through decoys or other measures, so that it misses its intended target.

[5] Reversible jamming means that the jamming does not damage the sensor, and that the sensor can resume normal operations once the jamming ends.

[6] An aerostat is a lighter-than-air object, such as a dirigible or balloon, that can stay stationary in the air.

[7] For further discussion, see P. Sprangle, J.R. Peñano, A. Ting, and B. Hafizi, "Propagation of High-Energy Lasers in a Maritime Atmosphere," *NRL Review 2004*. (Accessed online at http://www.nrl.navy.mil/research/nrl-review/2004/ featured-research/sprangle/.)

[8] Lasers being developed for potential shipboard use produce light with wavelengths in the near-infrared portion of the spectrum. Sweet spots in this part of the spectrum include wavelengths of 0.87 microns, 1.045 microns, 1.24 microns, 1.62 microns, 2.13 microns, and 2.2 microns. (Other sources, such as the research paper cited in footnote 7, cite somewhat different figures for sweet spot wavelengths, depending in part on whether sweet spot is for water vapor alone, or for multiple sources of atmospheric absorption and scattering.)

[9] The Navy installs multiple CIWS systems on certain ships not only to improve their ability to handle a saturation attack, but also to ensure that each ship has full (i.e., 360-degree CIWS)

coverage around the ship. A desire for 360- degree laser coverage could be another reason for installing multiple lasers on a ship.

[10] For a discussion of laser power levels, see "Required Laser Power Levels for Countering Targets."

[11] A March 2014 press report states, "Laser weapons like those developed by the United States pose little threat to [the Chinese military] ... because mainland [Chinese] researchers have pioneered coatings that can deflect beams and render them harmless, mainland scientists say." (Stephen Chen, "US Lasers? PLA Preparing To Raise Its Deflector Shields," *South China Morning Post* (*www.scmp.com*), March 10, 2014.) Another observer notes, "Lethality or desired levels of military effect are direct functions of the applied energy flux at the target and the 'race' between carriage of heat away from the material and bulk heating in a manner that causes the failure of the materials ... at high flux levels, surface ablation processes can create a dense outgassing cloud above the surface that absorbs the applied laser energy away from the surface thus acting to protect the target against vast power increases." (Email from James Kiessling, DT&E Space and Missile Defense Systems, March 13, 2014.)

[12] For more on the issue of collateral damage to satellites, see **Appendix J**.

[13] For an additional (and somewhat similar) discussion of the potential advantages and limitations of lasers, see Richard J. Dunn, III, *Operational Implications of Laser Weapons*, Northrop Grumman Analysis Center Papers, September 2005, pp. 10-12.

[14] Swarm boats are small, fast boats that attack a larger ship by operating in packs, or swarms, so as to present the larger ship with a complex situation of many hostile platforms that are moving rapidly around the ship in different directions.

[15] A laser with perfect BQ – meaning that the laser's light spot is focused to the physical diffraction limit – is said to have a BQ of 1.0. A beam that is focused to the physical diffraction limit is focused as well as the laws of nature allow. Lasers with the wavelengths considered in this report that are focused to the physical diffraction limit would, if fired in a vacuum, experience very little spreading out of the laser spot as the beam travels further and further from the source.

A BQ of 2.0 means that the laser's light spot at a given range is twice as large in diameter as an otherwise-same laser with a BQ of 1. The Navy considers a BQ of 1.1 to 5 to be high, and a BQ of 5.1 to 20 to be moderate. Achieving a BQ of 1 to 5 generally adds complexity and cost to the system. In general, the longer the range to the target, the more important BQ becomes.

[16] As discussed earlier, atmospheric absorption, scattering, and turbulence are affected by the laser's light wavelength and the use of adaptive optics.

[17] Jitter becomes more important as BQ improves and range increases.

[18] Some military lasers, such as the Air Force's Airborne laser (ABL), are chemically powered. Development work on potential shipboard lasers focuses on electrically powered lasers because such lasers can be powered by a ship's existing electrical power system, whereas a chemically powered laser would require the ship to be periodically resupplied with the chemicals used by the laser. Resupplying the ship with the chemicals could require the ship to temporarily remove itself from the battle. In addition, the Navy would need to establish a new logistics train to provide the chemicals to Navy surface ships, and loading and storing the chemicals on ships would create a handling risk for crew members, since the chemicals in question are toxic.

[19] As mentioned earlier the Phalanx CIWS is a radar-controlled Gatling gun that fires bursts of 20mm shells.

[20] Threat-representative means that the UAV is generally similar in design and capabilities to UAVs operated by potential adversaries.

[21] For a Navy press release about this test, see NAVSEA (Naval Sea Systems Command) press release dated May 28, 2010, and entitled "Navy Laser Destroys Unmanned Aerial Vehicle in a Maritime Environment," accessed online at http://www.navsea.navy.mil/PR2010 /PressRelease_20100528_Laser%20Destroys%20UAV.pdf. The UAVs engaged in these tests were BQM-147s, which various sources describe as low-cost, propeller-driven UAVs with a length of about 5 feet, a wingspan of about 8 feet, and a maximum speed of 100 knots or less.

[22] Mike McCarthy, "Navy Deploying Laser For Taking Out Drones," *Defense Daily*, April 9, 2013; Graham Warwick, "U.S. Navy Planning Gulf Deployment For Laser Weapon," *Aerospace Daily & Defense Report*, April 9, 2013: 6.

[23] Source: Navy information paper dated June 6, 2011, provided by the Navy to CRS and CBO on June 14, 2011. DOD uses TRL ratings to characterize the developmental status of many weapon technologies. DOD TRL ratings range from 1 (basic principles observed and reported) to 9 (actual system proven through successful mission operations). For the definitions of all 9 DOD TRL ratings, see Appendix II.

[24] The $17 million figure was provided in a Navy briefing to CRS. A May 11, 2010, press report quoted a Navy official as estimating the cost at $15 million:

"I think the total system, when we finally get it out there, will be on the order of $15 million per system and then there will be no ordnance costs, no logistics tail for maintaining the ordnance, no depots to overhaul ordnance, and no fire suppression as you move this ordnance around," [Capt. Dave Kiel, Naval Sea Systems Command (NAVSEA) directed energy and electric weapons program manager] said.

(Geoff Fein, "Navy Leveraging Commercial Lasers To Shoot Down UAVs," *Defense Daily*, May 11, 2010: 3-4.)

[25] Carlo Munoz, "New Laser-Based Weapon For Navy Fleet Protection Operations In The Works," *Defense Daily*, April 11, 2011. See also Marc Selinger, "Lasers on the High Seas," http://www.boeing.com, November 28, 2011, accessed November 28, 2011, at http://www.boeing.com/Features/2011/11/bds_tls_11_28_11.html.

[26] BAE Systems press release dated April 7, 2011, entitled "BAE Systems Selected to Demonstrate Tactical Laser System for the U.S. Navy;" Carlo Munoz, "New Laser-Based Weapon For Navy Fleet Protection Operations In The Works," *Defense Daily*, April 11, 2011.

[27] Megan Eckstein, "FEL Looks Good At CDR, But Project Halted In Favor of SSL Development," *Inside the Navy*, March 26, 2012.

[28] Megan Eckstein, "BAE, Boeing Pushing To Get Tactical Laser System To At-Sea Testing," *Inside the Navy*, January 28, 2013.

[29] See Northrop Grumman press release dated July 26, 2010, and entitled "Northrop Grumman-Built Maritime Laser Demonstration System Proves Key Capabilities for Shipboard Operations, Weaponization," accessed online at http://www.irconnect.com/noc/press/pages /news_releases.html?d=197321.

[30] See Northrop Grumman press release dated September 30, 2010, and entitled "Northrop Grumman-Built Maritime Laser Demonstration System Shows Higher Lethality, Longer Ranges at Potomac River Test Range; U.S. Navy Solid-State Laser's Mature Technology is Ready for Marine Environment;" accessed online at http://www.irconnect.com/ noc/press/pages/news_releases.html?d=202703.

[31] Andrew Burt, "Navy's First At-Sea Maritime laser Weapon Test Encounters Delays," *Inside the Navy*, November 15, 2010.

[32] Geoff S. Fein, "MLD Test Moves Navy a Step Closer to Lasers for Ship Self-Defense," April 8, 2011 (Office of Naval Research news release, accessed online at http://www.onr.navy.mil/en/Media-Center/Press-Releases/2011/ Maritime-Laser-MLD-Test.aspx.)

[33] Graham Warwick, "Northrop To Offer High-Power Ship Laser Within Four Years," *Aerospace Daily & Defense Report*, May 16, 2011: 4.

[34] Source: Navy information paper dated December 3, 2010, provided by the Navy to CRS on December 3, 2010. As mentioned in footnote 23, DOD uses TRL ratings to characterize the developmental status of many weapon technologies. DOD TRL ratings range from 1 (basic principles observed and reported) to 9 (actual system proven through successful mission operations). For the definitions of all 9 DOD TRL ratings, see Appendix II.

[35] For a description of INPs, see Appendix G.

[36] A low power Terahertz Sensor FEL is also being developed under the INP, with a prototype scheduled to be available in FY2015. ONR states that "Possible uses of this system include [target] interrogation, sensing and discrimination of high value targets, and weapons of mass destruction detection."

[37] Megan Eckstein, "FEL Looks Good At CDR, But Project Halted In Favor of SSL Development," *Inside the Navy*, March 26, 2012.

[38] Source: Navy information paper dated December 3, 2010, provided by the Navy to CRS on December 3, 2010. As mentioned in footnote 23, DOD uses TRL ratings to characterize the developmental status of many weapon technologies. DOD TRL ratings range from 1 (basic principles observed and reported) to 9 (actual system proven through successful mission operations). For the definitions of all 9 DOD TRL ratings, see Appendix H.

[39] The term non-material efforts refers to actions other than the acquisition of new or modernized equipment, such as making changes in doctrine or tactics.

[40] Memorandum dated December 12, 2011, from the Under Secretary of the Navy, to various Navy offices, on the subject: "Naval Directed Energy Steering Group Charter," posted at InsideDefense.com (subscription required) June 18, 2012. See also: Megan Eckstein, "Naval Directed-Energy Steering Group Outlining Future Of DE Weapons," *Inside the Navy*, June 15, 2012.

[41] Department of the Navy, *A Directed Energy Vision for U.S. Naval Forces*, 2 pp., provided to CRS by Navy Office of Legislative Affairs, August 20, 2012. Emphasis as in original.

[42] Megan Eckstein, "Directed-Energy Roadmap Due This Fall, Will Begin Guiding Budgets," *Inside the Navy*, August 5, 2013.

[43] Megan Eckstein, "ONR Planning First Solid-State Laser Weapon Prototypes On DDG, LCS," *Inside the Navy*, August 20, 2012. Ellipse in the quote as in the article.

[44] "Navy Leaders Announce Plans for Deploying Cost-Saving Laser Technology," *Navy News Service*, April 8, 2013; Thom Shanker, "Navy Deploying Laser Weapon Prototype Near Iran," *New York Times*, April 9, 2013: 4; Mike McCarthy, "Navy Deploying Laser For Taking Out Drones," *Defense Daily*, April 9, 2013; Graham Warwick, "U.S. Navy Planning Gulf Deployment For Laser Weapon," *Aerospace Daily & Defense Report*, April 9, 2013: 6; Megan Eckstein, "Navy-Built Laser Weapon System Will Begin Demo On Ponce In Early 2014," *Inside the Navy*, April 15, 2013. See also Office of Naval Research, "All Systems Go: Navy's Laser Weapon Ready for Summer Deployment," Navy News Service, April 7, 2014.

[45] Lara Seligman, "Navy-built LaWS To Begin Demo This Summer, IOC Slated For FY-20-21," *Inside the Navy*, March 24, 2014.

[46] Office of Naval Research news release, "Historic Leap: Navy Shipboard Laser Operates in Persian Gulf," December 10, 2014, accessed December 23, 2014, at http://www.onr.navy.mil/en/Media-Center/Press-Releases/2014/LaWSshipboard-laser-uss-ponce.aspx. The news release was also posted as David Smalley, "Historic Leap: Navy Shipboard Laser Operates in Arabian Gulf," *Navy News Service*, December 10, 2014, accessed December 23, 2014, at http://www.navy.mil/submit/display.asp?story_id=84805. See also Dan Lamothe, "With Photos And Video, Navy Shows How Its New Laser Gun Works At Sea," *Washington Post (www.washingtonpost.com)*, December 10, 2014; David Larter, "Navy's First Laser Gun Shines In Deployed Exercises," Defense News (www.defensenews.com), December 11, 2014; Julian E. Barnes, "Navy Tests Laser Weapon on Drones, Boats," Wall Street Journal (www.wsj.com), December 9, 2014; Hendrick Simoes, "Navy Encouraged By Performance of Laser System on USS Ponce," *Stars and Stripes (www.stripes.com)*, December 6, 2014; Tony Capaccio, "U.S. Navy Deploys Its First Laser Weapon in the Persian Gulf," *Bloomberg News (www.bloomberg.com)*, November 14, 2014; Lara Seligman, "Navy Fires Laser Weapon For First Time From Forward-Deployed Vessel," *Inside the Navy*, October 76, 2014.

[47] Mike McCarthy, "Navy Authorized To Use Ship-Based Laser In Battle," *Defense Daily*, December 11, 2014: 3. See also Sam LaGrone, "U.S. Navy Allowed to Use Persian Gulf Laser for Defense," *USNI News (*http://news.usni.org*)*, December 10, 2014; Philip Ewing, "Navy Declares Laser Weapon 'Operational,'" *Politico Pro (Pro Defense Report)*, December 10, 2014.

[48] Lara Seligman, "Navy-built LaWS To Begin Demo This Summer, IOC Slated For FY-20-21," *Inside the Navy*, March 24, 2014.

[49] Transcript of hearing.

[50] CHAMP stands for Counter-Electronics High Power Microwave Advanced Missile Project, another DOD effort.

[51] Transcript of hearing. See also Lara Seligman, "ONR Chief: If Ponce Demo A Success, LaWS Will Deploy On Other Ships," *Inside the Navy*, March 31, 2014; John C. Marcario, "Navy's Laser Weapon System Facing Big Test," *Seapower* (*www.seapowermagazine.org*), March 26, 2014.

[52] *Department of Defense Fiscal Year (FY) 2016 President's Budget Submission, Navy Justification Book Volume 1 of 5, Research, Development, Test & Evaluation, Navy, Budget Activities 1, 2 & 3*, February 2015, p. 89.

[53] *Department of Defense Fiscal Year (FY) 2016 President's Budget Submission, Navy Justification Book Volume 1 of 5, Research, Development, Test & Evaluation, Navy, Budget Activities 1, 2 & 3*, February 2015, p. 89.

[54] *Department of Defense Fiscal Year (FY) 2016 President's Budget Submission, Navy Justification Book Volume 1 of 5, Research, Development, Test & Evaluation, Navy, Budget Activities 1, 2 & 3*, February 2015, p. 328.

[55] *Department of Defense Fiscal Year (FY) 2016 President's Budget Submission, Navy Justification Book Volume 2 of 5, Research, Development, Test & Evaluation, Navy Budget Activity 4*, February 2015, p. 906.

[56] Free space optics are those arranged so that the light travels from one optical element (such as a mirror) to another, with an air gap (i.e., free space) in between.

[57] LRUs are sealed, box-like containers enclosing many of a weapon's components. LRUs support a modular approach to maintenance in which personnel repair the weapon by removing a faulty LRU and replacing it with another.

[58] For more on the Flight III DDG-51, see CRS Report RL32109, *Navy DDG-51 and DDG-1000 Destroyer Programs: Background and Issues for Congress*, by Ronald O'Rourke.

[59] Grace V. Jean, "Navy Aiming for Laser Weapons at Sea," National Defense, August 2010, accessed online at http://www.nationaldefensemagazine.org/archive/2010/August /Pages/NavyAimingforLaserWeaponsatSea.aspx.

[60] The issue of thermal blooming in "down-the-throat" engagements is of particular concern for a megawatt-class laser. Since carriers and amphibious assault ships are potential high-value targets for an attacker, it might make more operational sense to install megawatt-class FELs on ships other than carriers or amphibious assault ships, so that those other ships could use their FELs to counter targets that are flying a crossing path toward a carrier or amphibious assault ship.

[61] For more on the CG(X) program, see CRS Report RL34179, *Navy CG(X) Cruiser Program: Background for Congress*, by Ronald O'Rourke.

[62] For more on the option of a new-design destroyer, see CRS Report RL32109, *Navy DDG-51 and DDG-1000 Destroyer Programs: Background and Issues for Congress*, by Ronald O'Rourke.

[63] For more on the CVN-78 program, see CRS Report RS20643, *Navy Ford (CVN-78) Class Aircraft Carrier Program: Background and Issues for Congress*, by Ronald O'Rourke.

[64] Michael Fabey, "U.S. Navy Beefing Up Shipboard Power For Laser Weapon Needs," *Aerospace Daily & Defense Report*, April 29, 2013: 4.

[65] This option could take the form of a provision broadly similar to Section 220 of the FY2001 defense authorization act (H.R. 4205/P.L. 106-398 of October 30, 2000), which set goals for the deployment of unmanned combat aircraft and unmanned combat vehicles. For the text of Section 220, see Appendix K.

[66] *Department of Defense Fiscal Year (FY) 2016 President's Budget Submission, Navy Justification Book Volume 1 of 5, Research, Development, Test & Evaluation, Navy, Budget Activities 1, 2 & 3*, February 2015, p. 89.

[67] *Department of Defense Fiscal Year (FY) 2016 President's Budget Submission, Navy Justification Book Volume 1 of 5, Research, Development, Test & Evaluation, Navy, Budget Activities 1, 2 & 3*, February 2015, p. 89.

[68] *Department of Defense Fiscal Year (FY) 2016 President's Budget Submission, Navy Justification Book Volume 1 of 5, Research, Development, Test & Evaluation, Navy, Budget Activities 1, 2 & 3*, February 2015, p. 328.

[69] *Department of Defense Fiscal Year (FY) 2016 President's Budget Submission, Navy Justification Book Volume 2 of 5, Research, Development, Test & Evaluation, Navy Budget Activity 4*, February 2015, p. 906.

[70] PMS-405 means Project Manager, Shipbuilding, office code 405.

[71] A June 6, 2010, press report states that "The system uses six commercial off-the-shelf five-and-a-half kilowatt welding lasers.... " (Dan Taylor, "Navy Testing Developmental Laser Against Small Surface Vessels," Inside the Navy, June 7, 2010.) Another source puts the total power of LaWS at 32 kW. (Larry Greenemeier, "U.S. Navy Laser Weapon Shoots Down Drones in Test, ScientificAmerican.com, July 19, 2010, accessed online at http://www.scientificamerican.com/article.cfm?id=laser-downs-uavs.)

[72] Other firms involved in the LaWS effort include IPG Photonics (the maker of the fiber SSLs), L-3 Communications, and Boeing. The LaWS effort also involves the Pennsylvania State

University Electro-Optics Center and the Johns Hopkins University Applied Physics Laboratory.

[73] Source: Navy information paper dated June 6, 2011, provided by the Navy to CRS and CBO on June 14, 2011.

[74] Navy information paper dated June 10, 2011, provided by the Navy to CRS and CBO June 22, 2011.

[75] See Northrop Grumman press release dated March 18, 2009, and entitled "Northrop Grumman Scales New Heights in Electric Laser Power, Achieves 100 Kilowatts From a Solid-State Laser," accessed online at http://www.irconnect.com/noc/press/pages/news_releases.html?d=161575.

[76] Geoff S. Fein, "MLD Test Moves Navy a Step Closer to Lasers for Ship Self-Defense," April 8, 2011 (Office of Naval Research news release, accessed online at http://www.onr.navy.mil/en/Media-Center/Press-Releases/2011/ Maritime-Laser-MLD-Test.aspx).

[77] Source: Navy information paper dated June 1, 2011, provided by the Navy to CRS and CBO on June 14, 2011.

[78] A 2004 media advisory from the Office of Naval research states:

In the FEL, electrons are stripped from their atoms and then whipped up to high energies by a linear accelerator. From there, they are steered into a wiggler—a device that uses an electromagnetic field to shake the electrons, forcing them to release some of their energy in the form of photons. As in a conventional laser, the photons are bounced between two mirrors and then emitted as a coherent beam of light. However, FEL operators can adjust the wavelength of the laser's emitted light by increasing or decreasing the energies of the electrons in the accelerator or the amount of shaking in the wiggler.

(Office of Naval Research media advisory released July 30, 2004, and entitled "Free-Electron Laser Reaches 10 Kilowatts," accessed online at http://www.onr.navy.mil/Media-Center/Press-Releases/ 2004/Free-Electron-Laser-10-Kilowatts.aspx.)

[79] These include the Naval Postgraduate School in California, the U.S. Naval Academy in Maryland, NRL, NSWC Carderock in Maryland, the Naval Air Weapons Center (NAWC) China Lake in California, NSWCDD, PMS405, and the Naval Warfare Systems Center Pacific (SPAWAR) in California.

[80] These are the Thomas Jefferson National Laboratory in Virginia, the Los Alamos National Laboratory in New Mexico, the Brookhaven National Laboratory in New York, and the Argonne National Laboratory in Illinois.

[81] These include the MIT Lincoln Laboratory in Massachusetts, Vanderbuilt University in Tennessee, Colorado State University, the University of California, the University of Wisconsin, Stanford University in California, Yale University in Connecticut, the University of Texas, and the University of Maryland.

[82] See Department of Defense contract announcement No. 804-10, dated September 7, 2010, accessed online at http://www.defense.gov/contracts/contract.aspx?contractid=4361. See also Geoff Fein, "ONR Awards Boeing $23 Million To Finish Free Electron Laser Design," Defense Daily, September 17, 2010: 3-4.

[83] Rob Anastasio, "Office of Naval Research Achieves Milestone in Free Electron Laser Program," *Navy News Service*, January 20, 2011.

[84] Source: Navy information paper dated June 1, 2011, provided by the Navy to CRS and CBO on June 14, 2011.

[85] Mike McCarthy, "Navy's Free Electron Laser Undergoing Design Review," *Defense Daily*, March 21, 2012: 7.

[86] Megan Eckstein, "FEL Looks Good At CDR, But Project Halted In Favor of SSL Development," *Inside the Navy*, March 26, 2012.

[87] Source: Navy information paper on directed energy dated August 26, 2010.

[88] Source: Department of Defense, *Technology Readiness Assessment (TRA) Deskbook*, July 2009, accessed online at http://www.dod.mil/ddre/doc/DoD_TRA_July_ 2009_ Read_Version.pdf.

[89] Treaties and Other International Acts Series 09-721.2, *Weapons, Blinding Laser Weapons (Protocol IV), Protocol Between the United States of America and Other Governments to the Convention on Prohibitions or Restrictions on the Use of Certain Conventional Weapons which may be deemed to be Excessively Injurious or to have Indiscriminate Effects of October 10, 1980*, accessed March 27, 2013, at http://www.state.gov/documents /organization/190580.pdf. See also U.S. Department of State, *Treaties in Force, A List of Treaties and Other International Agreements of the United States in Force on January 1, 2012*, page 483.

[90] Department of Defense, *CCW: Article by Article Analysis of the Protocol on Blinding Laser Weapons*, accessed online at http://www.acq.osd.mil/tc/treaties/ccwapl/artbyart_pro4.htm. In January 1997, Secretary of Defense William Perry issued a memorandum regarding DOD policy on blinding lasers which states in its entirety:

The Department of Defense prohibits the use of lasers specifically designed to cause permanent blindness and supports negotiations to prohibit the use of such weapons. However, laser systems are absolutely vital to our modern military. Among other things, they are currently used for detection, targeting, range-finding, communications, and target destruction. They provide a critical technological edge to U.S. forces and allow our forces to fight, win and survive on an increasingly lethal battlefield. In addition, lasers provide significant humanitarian benefits. They allow weapon systems to be increasingly discriminate, thereby reducing collateral damage to civilian lives and property. The Department of Defense recognizes that accidental or incidental eye injuries may occur on the battlefield as the result of the use of lasers not specifically designed to cause permanent blindness. Therefore, we continue to strive, through training and doctrine, to minimize these injuries.

(Memorandum dated January 17, 1997, from Secretary of Defense William J. Perry to the secretaries of the military departments, et al, on DOD policy on blinding lasers, provided to CRS on October 4, 2010, by the Navy Office of Legislative Affairs.)

Paragraph 4.3 of DOD Instruction 3100.11 of March 31, 2000, on the illumination of objects in space by lasers, states: "The use of lasers specifically designed to cause permanent blindness in humans is prohibited, in accordance with [the above-cited January 17, 1997, memorandum from the Secretary of Defense]."

[91] Department of Defense, *CCW: Article by Article Analysis of the Protocol on Blinding Laser Weapons*, accessed online at http://www.acq.osd.mil/tc/treaties/ccwapl/artbyart_pro4.htm.

[92] Department of Defense, *CCW: Article by Article Analysis of the Protocol on Blinding Laser Weapons*, accessed online at http://www.acq.osd.mil/tc/treaties/ccwapl/artbyart_pro4.htm.

[93] Department of Defense, *CCW: Article by Article Analysis of the Protocol on Blinding Laser Weapons*, accessed online at http://www.acq.osd.mil/tc/treaties/ccwapl/artbyart_pro4.htm.

[94] The text of the 1997 Secretary of Defense memorandum quoted in footnote 90 is slightly different at this point. Instead of "legitimate laser systems," the 1997 memorandum uses the phrase "lasers not specifically designed to cause permanent blindness."

[95] *[Report of] Defense Science Board Task Force on Directed Energy Weapons*, Washington, December 2007, pp. xiixiii. Ellipsis and material in brackets as in original.

[96] Department of Defense Instruction Number 3100.11, March 31, 2000, on Illumination of Objects in Space by lasers, paragraph 4.2.

In: Navy Shipboard Lasers
Editor: William C. Jenkins
ISBN: 978-1-63483-372-1

Chapter 2

PROPAGATION OF HIGH ENERGY LASER BEAMS THROUGH ATMOSPHERIC STAGNATION ZONES*

Joseph R. Penano, Phillip Sprangle and Bahman Hafizi

ABSTRACT

Stagnation zones, i.e., regions in which the effective transverse wind velocity is zero, can greatly enhance the thermal blooming of high energy laser beams (HELs) in the atmosphere. An approximate expression for the Strehl ratio of a focused HEL beam propagating through a stagnant zone is derived. The propagation of a HEL beam in a maritime atmosphere is modeled in a fully three-dimensional and time-dependent manner using the HELCAP propagation code. In the simulations, the beam is focused onto a remote target and a stagnation zone is created by slewing the laser in the direction of the wind. The laser power delivered to the target is calculated as a function of slew rate. For the parameters considered, it is found that a stagnation zone near the laser source has little effect on the propagation efficiency while a stagnation zone near the target can significantly reduce the power on the target.

* This is an edited, reformatted and augmented version of a report, NRL/MR/6790--06-8925, issued by the Naval Research Laboratory, January 27, 2006.

I. INTRODUCTION

High energy laser beams propagating through the atmosphere can be severely defocused or deflected by thermal blooming [1].The thermal blooming process is driven by a small fraction of the laser energy that is absorbed by the molecular and aerosol constituents of air [2, 3]. The absorbed energy locally heats the air and leads to a decrease in the air density which modifies the refractive index. The refractive index variation leads to a defocusing or deflection of the laser beam. In the presence of a transverse wind, the region of heated air is convected out of the beam path and a steady-state situation is realized [1]. In general, however, the intensity of a beam undergoing thermal blooming is a function of both time and spatial position, particularly in a stagnation zone, where the effective wind velocity is zero.

Stagnation zones are particularly detrimental to HEL propagation since, without an effective clearing mechanism for the heated air, the strength of the thermal lens grows in time. In this situation, the defocusing process is eventually limited by thermal conduction or buoyancy. However, by the time these processes become effective, the beam may have already been severely degraded.

Thermal blooming in the presence of a stagnation zone was experimentally observed in a laboratory experiment using a 10W, CO2 laser passed through an absorption cell containing CO_2 gas. The cell was pivoted to simulate a wind profile containing a stagnation zone. This experiment was also modeled using a code that solved time-dependent thermal blooming equations in the isobaric regime [4].

In this study we examine the effect of a stagnation zone on a HEL beam focused onto a remote target in a maritime atmosphere. We first analyze the effect of a stagnation zone on the propagation of a focused beam and calculate an approximate expression for the relative intensity (Strehl ratio) on target as a function of focusing geometry and stagnation zone position. We then utilize the HELCAP code [3], a 3D, time-dependent numerical simulation, to examine the propagation of a megawatt-class HEL beam through a maritime atmosphere which contains a stationary stagnation zone along the laser path. The laser target configuration used in these simulations is shown in Figure 1. Our simulation model includes time-dependent thermal blooming as well as aerosol and molecular absorption [2, 3], turbulence [5], aerosol and molecular scattering [6], and aerosol heating and vaporization [7-10].

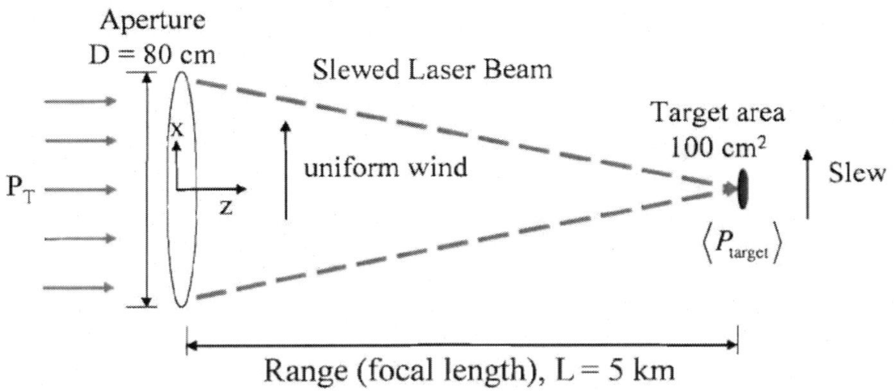

Figure 1. Schematic diagram showing the simulation geometry. A HEL beam of power P_T is focused at a range of L = 5 km in the presence of a uniform wind with speed Vw = 5 mr/sec. The beam is slewed in the direction of the wind to create a stagnation zone along the propagation path.

We use the simulations to calculate the laser power delivered to a remote target for different wind profiles which place the stagnation zone at different locations along the laser propagation path.

II. Analysis of HEL Propagation Through a Stagnation Zone

The effect of a stagnation zone on HEL propagation in the atmosphere is highly dependent on laser and atmospheric parameters as well as on the propagation geometry. In general, the problem requires a numerical simulation to determine the amount of power reaching a remote target. Before presenting results of full-scale propagation simulations, however, it is useful to analyze the following simpler problem.

Consider a focused laser beam of wavelength, λ, propagating from *z* = 0 to a remote target at a range *z* = *L,* where *z* denotes the propagation coordinate. The focal length of the laser beam, L_t, is such that L_l *L.* The wind speed, *V(z)* is only a function of z and has a stagnation point defined by $V(z_o)$ = 0, where 0 < z_o < *L.* The atmosphere is characterized by an absorption coefficient, *a* , and a scattering coefficient, fi α. Turbulence is neglected in this analysis. This configuration allows an approximate expression for the Strehl ratio (relative intensity) at the target to be calculated analytically as a function of both the location of the absorbing region and the laser focal length. The propagation of the laser beam is described by the paraxial wave equation

$$\frac{\partial A}{\partial z} = \left(\frac{i}{2k_o}\nabla_\perp^2 + i\frac{\omega_o}{c}\delta n_{TB} - \frac{\beta}{2} \right) A \tag{1}$$

where the laser electric field is E = *A* exp[*i(co, z/c — co, t)] e$_x$ /2 + c.c.*, *co,* = 27z¯c / , e$_Y$ is a unit polarization vector in the x direction, k$_o$ = n$_o$ *co, /c* , n$_o$ is the ambient refractive index, v1 is the gradient operator in the transverse direction, *c* is the speed of light in vacuum, *8n$_{T8}$* is the change in refractive index induced by the laser, and *c.c.* denotes the complex conjugate. In the absence of a transverse wind or slew, there is axial symmetry about the z-axis and all spatially varying quantities are functions of *(r,z,t)* only, where *r* is the radial coordinate and *I* is time. Writing A(r, z, t) = B(r,z,t)exp[i8(r,z,t)], where B , 6 , and 8 n$_7$ are assumed to be real, leads to the following equations for the laser intensity *I* (— B^2) and the transverse wavenumber k$_i$ = V$_1$0:

$$\frac{\partial I}{\partial z} = -\frac{1}{k_o}\nabla_\perp\cdot(\mathbf{k}_\perp I) - \beta I \tag{2}$$

$$\frac{\partial \mathbf{k}_\perp}{\partial z} = \frac{1}{8k_o}\nabla_\perp\left\{\frac{1}{I^2}\left(2I\nabla_\perp^2 I - (\nabla_\perp I)^2\right)\right\} - \frac{1}{k_o}\mathbf{k}_\perp\cdot\nabla_\perp\mathbf{k}_\perp + \frac{\omega_o}{c}\nabla_\perp\delta n_{TB} \tag{3}$$

On the right side of Eq. (3), the first term is of order ~ /*R^3* where *R* is the laser spot size. The second term is of order ~ /*R(L^2_fλ)* for a focused beam and the third term is of order ~ δn$_{TB}$/ *(λA,)*: We consider parameters such that *R* ~ 50cm, λ ~ 1μm , and *L$_f$* — 5 km. In this parameter regime, the second term is much larger than the first term.

The third term is much larger than the second term for $\delta n_{TB} > 10^{-8}$, which is typically encountered within a stagnation zone. Hence, for the parameter regime of interest, we neglect the first and second terms on the right side of Eq. (3) and integrate to solve for k$_1$, i.e.,

$$\mathbf{k}_{\perp}(r,z,t) \approx \mathbf{k}_{\perp 0}(r) + \frac{\omega_o}{c}\int_0^z \nabla_{\perp}\delta n_{TB}(r,z',t)\,dz' \tag{4}$$

where $k_{10}(r)$ is the initial transverse wavenumber at $z = 0$. Equation (4) represents k_1 in the geometric optics limit.

Equation (2) can be integrated to yield

$$I(r,z,t) = I(r,0,t)e^{-\beta z}\exp\left\{-\frac{c}{\omega_0}\int_0^z \frac{1}{I(r,z',t)}\nabla_{\perp}\cdot\left[\mathbf{k}_{\perp}(r,z',t)I(r,z',t)\right]dz'\right\} \tag{5}$$

which is an integral equation for $I(r,z,t)$ that can be solved approximately by replacing $I(r,z,t)$ in the right side by $I_0(r,z)$, i.e., the unperturbed intensity for a homogeneous medium $(\delta n_{TB} = 0)$. The unperturbed intensity satisfies

$$I_0(r,z) = I(r,0)e^{-\beta z}\exp\left\{-\frac{c}{\omega_0}\int_0^z \frac{1}{I_0(r,z')}\nabla_{\perp}\cdot\left[\mathbf{k}_{\perp 0}(r)I_0(r,z')\right]dz'\right\} \tag{6}$$

Substituting Eq. (4) into Eq. (5) yields the relative intensity [11]

$$\frac{I(r,z,t)}{I_0(r,z)} = \exp\left\{-\int_0^z \frac{dz'}{I_0(r,z')}\nabla_{\perp}\cdot\left[I_0(r,z')\int_0^{z'} dz''\,\nabla_{\perp}\delta n_{TB}(r,z'',t)\right]\right\} \tag{7}$$

The refractive index change due to thermal blooming is given by $\delta nTB = (n_o - 1)\delta\rho/\rho_0$, where ρ_o is the mass density, and $\delta\rho$ is the perturbed mass density. In the isobaric regime, the mass density change is given by [1, 3]

$$C_p T_o\left(\frac{\partial}{\partial t} + \mathbf{V}\cdot\nabla - \frac{\kappa}{C_p\rho_o}\nabla^2\right)\delta\rho = -\alpha_{TB}(\mathbf{r},t)I \tag{8}$$

where K is the thermal conductivity, V is the effective wind velocity, Cp is the specific heat at constant pressure, T_o is the ambient temperature, and I is the time-averaged laser intensity. In general, the absorption coefficient, α_{TB} contains both aerosol and molecular contributions and can be time dependent. For the simplified analysis of this Section, however, we take it to be constant in time. The isobaric approximation is valid for times greater than the hydrodynamic time R/C_s, where C_s is the acoustic speed. We consider time scales less than the thermal conduction time, $C_p \rho_0 R^2 / \kappa$, and neglect the thermal conduction term in Eq. (8).

For the purpose of obtaining an approximate closed form expression for Eq. (7), we solve Eq. (8) approximately by writing the operator V • V ≈ 1 $/r_e(z)$ where $r_e(z) = R(z)/|V(z)|$ is the clearing time. Using this approximation, we can integrate Eq. (8) to obtain

$$\delta n_{TB}(r,z,t) \approx -\gamma_{TB} \alpha I_0(r,z) \tau_c(z) \{1 - \exp[-t/\tau_c(z)]\} \quad 9)$$

where $Y_{TB} = (no -1)/(C_P T_o P_o)$.

The unperturbed intensity is taken to have a Gaussian transverse profile

$$I_0(r,z) = \frac{I_{00} R_0^2}{R^2(z)} \exp\left[-\frac{r^2}{R^2(z)}\right] e^{-\beta z} \quad (10)$$

where 1_{00} denotes the intensity on axis at z = 0. The spot size *R* is a function of z and the initial spot size is denoted by R_o=R(z = 0). Substituting Eqs. (9) and (10) into Eq. (7), results in an expression for the on-axis Strehl ratio at z = L i.e.,

$$\frac{I(r=0,L,t)}{I_0(r=0,L)} = \exp\left[-8 I_{00} R_0^2 \gamma_{TB} \alpha \int_0^L dz' \int_0^{z'} dz'' \, G(z'',t)\right] \quad (11)$$

where $G(z,t) = \tau_c(z)[1$ — exp(—t $/\tau_{c}$, (z))]exp(-β_z)/$R^4(z)$. To obtain an approximate analytic expression for the right side of Eq. (11), we note that for a weakly focused beam, the integrand G is sharply peaked and grows linearly with time in the vicinity of the stagnation zone, i.e., $G(z_o,t) = t$ exp(—β z_o)/ $R^4(z_o)$. Hence, we can approximate the integration over z'' in Eq. (11) by

$$\int_0^{z'} dz''\, G(z'',t) \approx G(z_0,t)\,\Delta z\,\Theta(z'-z_0) \tag{12}$$

where Δz is the characteristic width of the stagnation region and Θ is the Heaviside step function.

Using Eqs. (12) in Eq. (11) results in

$$\frac{I(r=0,L,t)}{I_0(r=0,L)} \approx \exp\left[-\frac{8\gamma_{TB}\alpha I_{00}L^2 t}{R_0^2}\left(\frac{\Delta z}{L}\right)\left(1-\frac{z_0}{L}\right)\frac{R_0^4\exp(-\beta z_0)}{R^4(z_0)}\right] \tag{13}$$

The width of the stagnation region, Δz , is in general a function of time and also the spatial variation of the wind profile. We can define the stagnation zone as the region where the condition $t/\tau_c(z) \ll 1$ is satisfied. In this case, the width of the stagnation zone can an be defined according to $t/\tau_c\,(z_o + \Delta z) = 1$.

We now consider specific analytic expressions for the wind profile $V(z)$ and laser spot size R(z). For a slewed laser beam, the wind profile can be written as

$$V(z) = V_{wind}(1 - z/z_0) \tag{14}$$

where V_{wind} is the ambient wind velocity, and the laser slew rate , Θ is related to z_0 according to $\Theta = V_{wind}/z_0$. Note that for this model, a stagnation zone at the transmitter (z = 0) implies an infinite slew rate. Assuming that the laser spot size does not vary appreciably within the stagnation zone,

$$\Delta z \approx z_0 R(z_0)/(V_{wind}\, t) \tag{15}$$

for the wind profile of Eq. (14).

For the linearly varying wind profile described above, Eq. (13) can be written as

$$\frac{I(r=0,L,t)}{I_0(r=0,L)} \approx \exp\left[-\frac{8\gamma_{TB}\alpha I_{00}L^2}{R_0 V_{wind}} f(z_0)\right] \tag{16}$$

where

$$f(z_0) = \left(\frac{z_0}{L}\right)\left(1-\frac{z_0}{L}\right)\frac{R_0^3 \exp(-\beta z_0)}{R^3(z_0)} \tag{17}$$

describes how the location of the stagnation zone affects the relative intensity on target. The larger the value of $f(z_0)$, the lower the intensity on the target.

We can derive an approximate functional form for $R(z)$ to be used in Eq. (17). Substituting Eq. (10) into Eq. (6), the laser spot size can be shown to satisfy $R(z) = R_0[1 - z(1-\eta)/L]$ for $z << L_f$, where the parameter $\eta = R(L)/R_o$, i.e., the ratio of the spot size on target to the initial spot size. In terms of the focal length, $\eta = 1 - (L/L_f)$. Phenomenologically, η can also parameterize focusing limitations due to beam quality or atmospheric turbulence.

Figure 2 plots the distortion function f versus z_o for various values of η. We assume a normalized scattering coefficient of $\beta L = 0.25$. For a beam that is focused near the target plane, the distortion at the target plane is largest when the stagnation zone is close to the target. In contrast, when the beam is collimated ($\eta \sim 1$), the distortion at the target plane is greater when the stagnation zone is closer to the laser source. Note that the case $z_0 = 0$, which corresponds to an infinite slew rate given the assumed wind profile of our model, is not physically realizable.

III. NUMERICAL SIMULATION

The propagation code used in this study is HELCAP (High Energy Laser Code for Atmospheric Propagation), which is a fully time-dependent, 3D code developed at the Naval Research Laboratory. HELCAP models the propagation of continuous and pulsed HEL beams through various atmospheric environments.

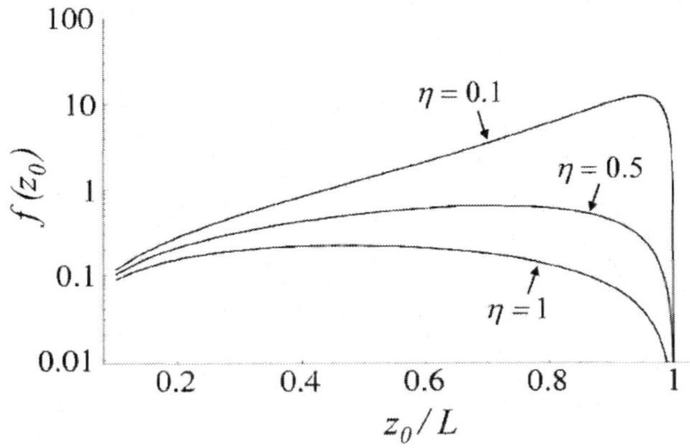

Figure 2. Distortion function *f* as defined by Eq. () plotted versus distance to stagnation zone zo for various values of the focusing parameter η = R(L) I R_0 where R(L) is the unperturbed laser spot size at the target plane (z = L), R_0 is the initial spot size at z = 0, and $\beta L = 0.25$.

It includes the effects of aerosol and molecular scattering, aerosol heating and vaporization, thermal blooming due to both aerosol and molecular absorption, and atmospheric turbulence. A more detailed description of the code can be found in Refs. [3] and [12].

HELCAP solves a nonlinear Schrodinger-like equation for the complex laser electric envelope, $A(x,y,z,t)$, which is of the form

$$\frac{\partial A}{\partial z} = \frac{ic}{2\omega_o}\nabla_\perp^2 A + \left[i\frac{\omega_o}{c}(\delta n_T + \delta n_{TB}) - \frac{1}{2}(\alpha_{TB} + \beta) \right] A + \sum_j S_j \tag{12}$$

where $A(x,y,z,t)$ is the complex laser electric field amplitude and the terms denoted by $\sum_j S_j$, represent other physical processes that are included in the code but are not important for the parameter regime considered here. The laser intensity is $I = c A A^* / 8\pi$. The absorption and scattering coefficients α_{TB} and β, respectively, contain both molecular and aerosol contributions. In general, they are time dependent due to aerosol heating and vaporization by the laser as described in Ref. [3]. The quantities δn_T and δn_{TB} denote the refractive index variation due to atmospheric turbulence and thermal blooming respectively. Atmospheric turbulence is modeled in the usual manner using phase screens for which the scale sizes of the index fluctuations are described by a Kolmogorov spectrum characterized by the parameter C_n^2 [5]. The mass density change due to thermal blooming is solved for according to Eq. (8), where the total absorption coefficient , α_{TB}, is time-dependent and contains both aerosol and molecular contributions in the manner described in Ref. [3].

For the numerical simulations presented here, we consider a slewed HEL beam focused at a range of L = 5km in the presence of a uniform transverse wind. The z-axis defines the direction of propagation and the wind and slew velocities are taken to be along the y-axis. The simulation geometry is shown in Figure 1. The high energy laser beam has an initial field profile given by $A = A_0 f(r) g(t) \exp(-r^2 / R_0^2)$, where $f(r) = \exp[-(2r/D)^{\ell}]$,$\ell = 20$, limits the transverse extent of the beam to the aperture diameter, D. The function $g(t)$ is the initial temporal profile of the beam which contains a pointing jitter characterized by an angular spread of 2 μrad and a white noise temporal spectrum. The transmitted power at the source is denoted by P_T.

The laser is propagated through a maritime environment in which the aerosol distribution is modeled using the Advanced Navy Aerosol Model (ANAM) [13] and the molecular absorption coefficient is calculated using MODTRAN4 [14]. The details of the atmospheric model and the parameters used to characterize a maritime environment are discussed in Ref. [3]. The ambient absorption and scattering coefficients for propagation at λ = 2.141 μm are a_{T8} = 6 x10' km-' and β= 0.05 km-' , respectively. In the simulations these quantities vary in space and time due to aerosol vaporization.

The effective wind speed along the propagation path is $V(z) = V_{wind} - \Theta z$ where z is the distance from the laser source, V_{wind} is the ambient wind speed, and Θ is the laser beam slew rate. For positive Θ, i.e., slewing in the direction of the wind, a stagnation zone is created at location $z_o = V_{wind} /$. In what follows, we investigate the effect of this stagnation zone on the propagation of a HEL beam for various slew rates which change the distance from the transmitter to the stagnation zone.

IV. Results

We consider a laser of wavelength λ= 2.141 μm and power P_T = 1.5 MW propagating through a maritime atmosphere with V_{wind} = 5 m / sec. The power delivered to a target of circular area 100 cm^2 at a range of 5 km is used as a figure of merit. In the simulations that follow, the target is always located such that the peak laser fluence at 5 km range is at the center of the target. The average power reaching the target is defined by

$$\langle P_{\text{target}} \rangle = \frac{1}{\tau_{dwell}} \int_0^{\tau_{dwell}} d\tau \int dxdy\, I(x, y, z = L, \tau) \tag{13}$$

where the dwell time $\tau_{dwell} = 1\text{ sec}$, and *dxdy* is the differential cross section which is integrated over the target area. The total laser energy reaching the target is $E_{\text{target}} = \langle P_{\text{target}} \rangle \tau_{dwell}$

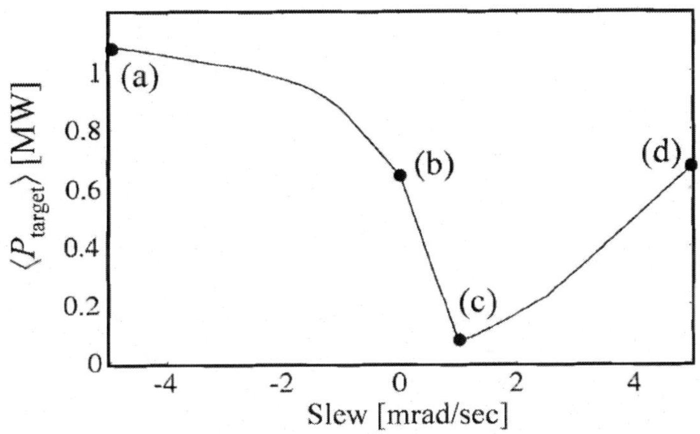

Figure 3. Average power on target $\langle P_{\text{target}} \rangle$, versus slew rate for $\lambda = 2.141\,\mu$rn, $P_T = 1.5$ MW , $V_{wind} = 5$ m / sec, $\tau_{dwell} = 1$ sec , and $L = 5$ km. Points (a), (b), (c), and (d) are labeled to correspond with fluence distributions shown in Figure 3.

Figure 3 plots $\langle P_{\text{target}} \rangle$ versus slew rate, 9, for the propagation configuration of Figure 1. A negative slew denotes that the slew direction is opposite to the wind direction. In this case, there is no stagnation zone between the laser and the target. A positive slew denotes slew in the direction of the wind. In this case, it is possible to have a stagnation zone somewhere along the propagation path. For example, a slew rate of 1 mradJsec places the stagnation zone at the target, while a slew rate of 5 mrad/sec places the stagnation zone 1 km from the laser source. Figure 3 shows that in the absence of slew (i.e., uniform wind), the power on target is $\langle P_{target} \rangle \approx 0.7$ MW and the propagation efficiency is ~ 50%. $\langle P_{target} \rangle$ increases relative to this value when the slew is opposite to the wind direction because the effectively larger wind speed along the entire propagation path mitigates thermal blooming. For a slew rate Θ = - 5 mrad / sec , for example, $\langle P_{target} \rangle \approx 1.1$ MW, which corresponds to a propagation efficiency of > 70%.

When the slew is increased in the direction of the wind, the power on target decreases sharply for Θ < 1 mrad /sec and reaches a minimum when Θ = 1 mrad /sec, i.e., when the stagnation zone is at the target. At this minimum, (P_{tatget}0.07 MW, which corresponds to a propagation efficiency of < 5%. Increasing the slew rate further brings the stagnation zone closer to the laser source, where the intensity is lower, and increases the effective wind velocity near the target. The result is that the power on target increases with slew rate. For a slew rate of Θ = 5 mrad/sec, $\langle P_{target} \rangle \approx 0.7$ MW. These results are consistent with the calculation in Sec. II, i.e., for a beam focused onto the target plane, a stagnation zone close to the target will result in the largest beam distortion.

Figure 4 plots the laser fluence on target calculated over a 1 sec dwell time. Panel labels, (a), (b), (c), and (d) correspond to the different slew rates indicated in Figure 3. Panel (a) corresponds to a slew rate Θ = -5 mrad/ sec , which gives the maximum power on target over the range of slew rates considered. The beam is well-focused and exhibits practically no thermal blooming effects. Panel (b) shows the beam profile when the slew rate is zero, i.e., a uniform wind profile. In this case the beam fluence profile shows the characteristic crescent shape associated with thermal blooming with the centroid of the beam deflected opposite to the direction of the wind.

Panel (c) shows the beam profile on target for the case when the stagnation zone is at the target. The beam is severely defocused as a result of thermal blooming. The peak in fluence near (x = 0, y = 0) is due to the leading edge of the beam (t < 20 msec), which is not affected by the time dependent thermal blooming process. Panel (d) shows the beam profile when the stagnation zone is 1 km from the target. In this case, the effective wind velocity near the target is in the opposite direction relative to that of panel (b). The value of $\left\langle P_{target} \right\rangle$ is similar to that of panel (b) but with the beam deflected in the opposite direction.

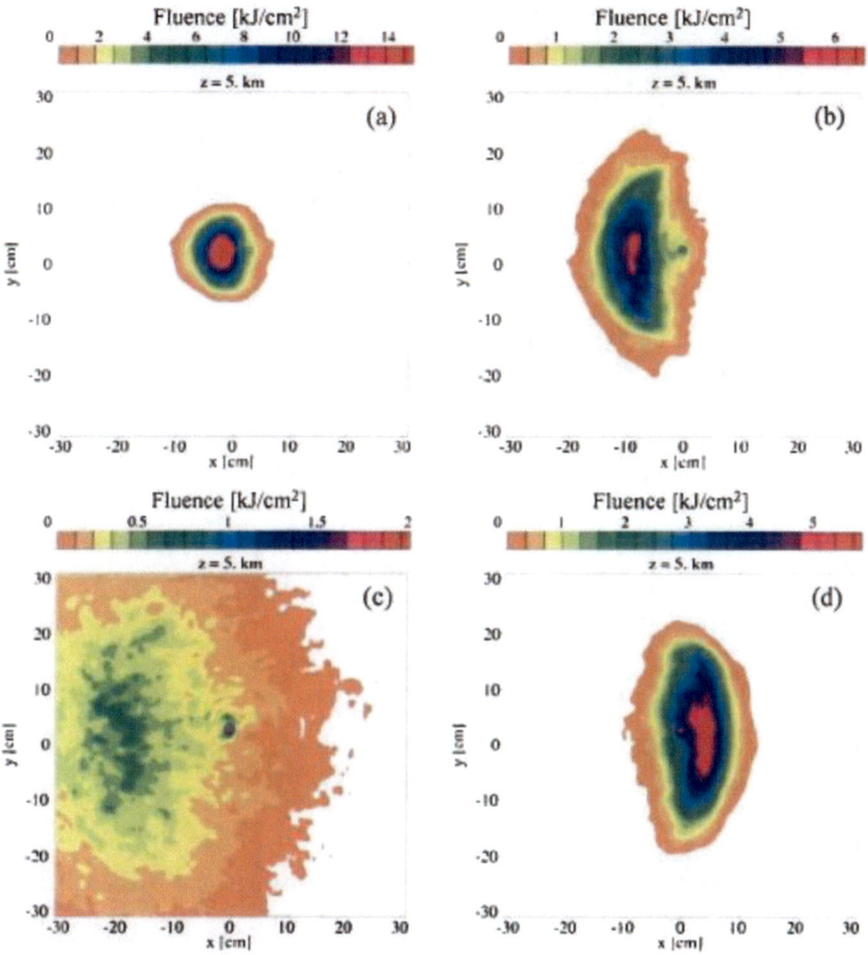

Figure 4. Laser fluence profiles at the target range for slew rates (a) $\Theta = -5$ mrad / sec., (b) $\Theta = 0$, (c) $0 = 1$ mrad/ sec , and (d) $\Theta = 5$ mrad /sec. Parameters are the same as those of Figure 2. Fluence is calculated over a 1 sec dwell time.

Figure 5 shows the time dependence of the beam intensity at the target. Figure 4a corresponds to a slew rate of $\Theta = -5$ mrad/ sec , when the power on target is maximum. In this case the beam is well-focused over the entire dwell time. The transverse profile is mainly affected by turbulence and pointing jitter. Figure 4b corresponds to a slew rate of $\Theta = 1$ mrad/sec , which results in the minimum power on target. For this case, it is seen that severe defocusing occurs for times > 20 msec, while the first ~ 10 msec of the beam is relatively unaffected by thermal blooming.

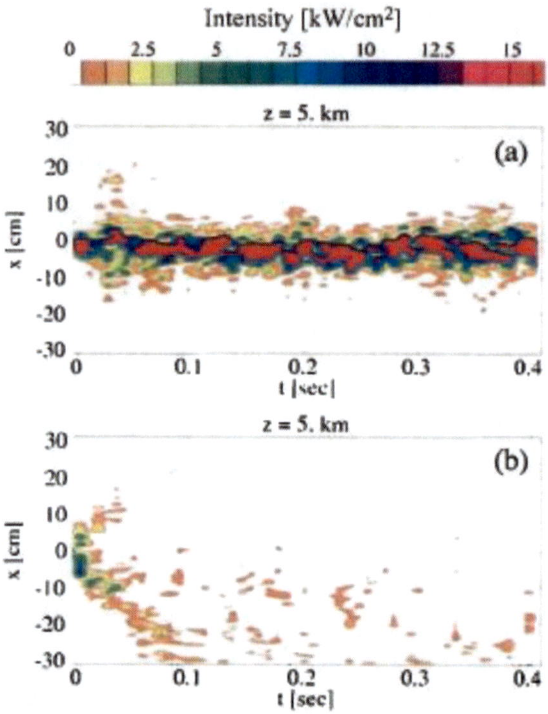

Figure 5. Laser intensity in the y = 0 plane versus transverse coordinate, x, and time, t, for slew rates (a) $\Theta = -5$ mrad / sec and (b) $\Theta = 1$ mrad / sec. Parameters correspond to those of Figure. 2.

Conclusion

Stagnation zones can significantly degrade the propagation efficiency of a high energy laser (HEL) through the atmosphere. We have analyzed the propagation of a focused HEL beam through a model atmosphere containing a stagnation zone. Our analysis shows that the amount of laser degradation caused by a stagnation zone is highly dependent on the focusing geometry of the laser beam and the location of the stagnation zone along the propagation path. For a collimated laser beam, a stagnation zone near the transmitter causes the largest distortion of the laser beam on the target. For a laser that is focused onto the target, a stagnation zone near the target causes the largest laser distortion and the fastest decrease in laser intensity. This finding is in agreement with the experimental results of Ref. [4].

We use a full scale numerical simulation (HELCAP) to model the propagation of a megawatt-class HEL beam through a realistic maritime atmosphere containing a stationary stagnation zone. The HEL beam is focused onto a remote target. The stagnation zone is created by slewing the HEL beam in the direction of the wind and location of the stagnation zone is varied by changing the slew rate. Consistent with our analysis, the simulations show that the power on target is minimized when the slew rate is such that the stagnation zone is located near the target plane. In this case, propagation efficiency can be reduced by an order of magnitude relative to an unslewed beam.

ACKNOWLEDGMENTS

This work was sponsored by the Joint Technology Office and Office of Naval Research.

REFERENCES

[1] Smith, D.C., "High-Power Laser Propagation - Thermal Blooming," *Proc. IEEE* 65, 1679 (1977).

[2] Brown R.T. and D.C. Smith, *J. App. Phys.* 46, 402 (1975).

[3] Spangle, P., J.R. Pefiano, B. Hafizi, "Optimum Wavelength and Power for Efficient Laser Propagation in Various Atmospheric Environments," NRUMR/6790-05- 8907.

[4] Berger, P.J., P.B. Ulrich, J.T. Ulrich, F.G. Gebhardt, *App. Opt.* 16, 345 (1977).

[5] The Infrared and Electro-Optical Systems Handbook, vol. 2, edited by F.G. Smith, Environmental Research Institute of Michigan, Ann Arbor, MI, and SPIE Optical Engineering Press, Bellingham, WA (1993).

[6] Measures, R.M., "Laser Remote Sensing, Fundamentals and Applications," Krieger Publishing, Malabar, FL (1992).

[7] Williams, F. A, *Int. J. Heat Mass Transfer*. 8, 575 (1965).

[8] Caledonia, G.E. and J.D. Teare, *J. Heat Transfer* 99, 281 (1977).

[9] Armstrong, R.L., Appl. Optics 23, 148 (1984); Armstrong, R.L., J. Appl. Phys. 56, 2142 (1984); Armstrong, R.L., S.A.W. Gerstl and A. Zardecki, *J. Opt. Soc. Am.* A 2, 1739 (1985).

[10] Davies, S.C. and J.R. Brock, App. Optics 26, 786 (1987).

[11] Gebhardt, F.G. and D.C. Smith, *IEEE J. Quant. Elec.* 7, 63 (1971); Buser, R.G., R.S. Rohde, P.J. Berger, F.G. Gebhardt, and D.C. Smith, *App. Opt.* 14, 2740 (1975).

[12] Sprangle, P., J.R. Pefiano and B. Hafizi, Phys. Rev. E 66, 046418 (2002); Pefiano, J.R., P. Sprangle, B. Hafizi, A. Ting, D.F. Gordon, C.A. Kapetanakos, *Phys. Plasmas* 11, 2865 (2004).

[13] van Eijk, A.M.J. and L.H. Cohen, "The ANAM-3.0 Development," TNO Physics and Electronics Laboratory, Jun. 2005; Piazzola, J., M.J. van Eijk, G. de Leeuw, Opt. Eng. 39, 1620 (2000); Zeisse, C.R., "NAM6: Batch Code fo the Navy Aerosol Model," Technical Report 1804, SPAWAR Systems Center, San Diego, CA 921525001, Oct. 1999.

[14] Berk, A., G.P. Anderson, P.K. Acharya, J.H. Chetwynd, L.S. Bernstein, E.P. Shettle, M.W. Matthew, and S.M. Adler-Golden, "MODTRAN4 User's Manual."

INDEX

A

accelerator, 51, 53, 54, 68
access, 18, 32, 48
adverse weather, 21
agencies, 19, 34
Air Force, 12, 13, 36, 45, 60, 61, 63
amplitude, 79
appropriations, 34
armed conflict, 57
armed forces, 36
assault, 29, 31, 67
assessment, 60
assets, 39
atmosphere, 6, 7, 71, 72, 73, 80, 84, 85
atoms, 68
attacker, 4, 67
authorities, 37
authority, 35
avoidance, 59
awareness, 15

B

ballistic missiles, 5, 8, 9, 10, 18, 27, 38, 62
barriers, 36
base, 3
beams, 41, 42, 45, 46, 52, 63, 71, 72, 78
benchmarks, 48
benefits, 58, 59, 69
blindness, 56, 57, 58, 69
boat, 8, 11, 13, 19, 20, 21, 45, 47, 48, 49

C

challenges, 6, 14, 15, 18, 19, 36
chemicals, 63
Chief of Staff, 62
China, 9, 11, 40, 47, 63, 68
classes, 20, 24
classification, 59
CO2, 72
coatings, 63
collaboration, 24, 47
collateral, 6, 8, 18, 57, 58, 59, 63, 69
collateral damage, 6, 8, 18, 58, 59, 63, 69
color, 41
commercial, 22, 27, 36, 37, 41, 45, 67
community, 55, 59
competition, 25
competitors, 26
complement, 49
complexity, 27, 28, 63
composition, 51
conduction, 72, 76
conductivity, 75
conference, 21
configuration, 72, 73, 81
conflict, 9
congress, v, 1, 2, 3, 25, 28, 30, 60, 67

conserving, 5
constituents, 72
construction, 28
containers, 67
convention, 56
conviction, 23
cooling, 5, 8, 15, 16, 18, 20, 28, 29, 41, 45, 47
cooperation, 36, 47
cooperative agreements, 34
cost, 2, 4, 6, 8, 11, 18, 19, 21, 22, 25, 26, 27, 28, 32, 33, 34, 36, 37, 52, 63, 64
crises, 35
cruise missiles, 5, 18, 38
crystalline, 46
cyberspace, 18
cycles, 36

D

danger, 35
data collection, 49
degradation, 85
demonstrations, 18, 20, 21
denial, 18
Department of Defense, vii, 1, 2, 33, 34, 35, 36, 57, 66, 67, 68, 69, 70
Department of Energy, 51
destruction, 21, 58, 69
detachment, 23
detection, 15, 58, 65, 69
deterrence, 48
diffraction, 63
diffusion, 19, 36
diodes, 40
discrimination, 57, 65
distribution, 45, 80
dominance, 36
DOT, 17
draft, 17, 42
drawing, 28

E

electric field, 74, 79
electricity, 4, 5, 30, 40
electromagnetic, 3, 7, 15, 18, 30, 32, 36, 68
electron(s), 1, 10, 51, 52, 53, 54, 68
employment, 18, 19, 56, 57, 59
energy, vii, 1, 2, 3, 12, 13, 14, 15, 16, 17, 18, 19, 20, 21, 22, 24, 28, 30, 31, 32, 33, 34, 35, 36, 37, 41, 45, 47, 48, 49, 51, 58, 59, 63, 64, 68, 69, 71, 72, 80, 81, 84
energy efficiency, 18
energy recovery, 51
engineering, 13, 15, 26, 42, 45, 61
environment(s), 11, 13, 14, 18, 22, 23, 26, 27, 32, 43, 48, 49, 52, 56, 78, 80
environmental conditions, 49
equipment, 28, 29, 51, 57, 65
erosion, 35
evidence, 52
exercises, 21
exploitation, 18, 55
exposure, 59

F

fabrication, 42, 52
fiber(s), 1, 10, 12, 25, 26, 27, 28, 40, 41, 46, 67
fire suppression, 64
fires, 4, 5, 63
fiscal year, 35, 37, 60, 61
flexibility, 52
fluctuations, 79
force, 37, 56, 58, 60, 61, 62
Ford, 29, 31, 67
funding, 2, 3, 12, 24, 25, 30, 31, 32, 33, 34, 60
funds, 33, 35

G

geometry, 72, 73, 80, 85
guidelines, 33

guiding principles, 17

H

handoff, 45
Hezbollah, 9
high winds, 21
House, 22, 32, 34, 37
House of Representatives, 34
human, 26, 27
humidity, 21, 27

I

ideal, 22
identification, 60
illumination, 69
images, 53
improvements, 41
incidence, 57
industrial environments, 27
industry, 3, 9, 10, 14, 20, 21, 37, 39, 47
inevitability, 57
injuries, 57, 58, 69
injury, 57
institutions, 51
integration, 18, 19, 42, 45, 49, 76
integrity, 35
intelligence, 18, 33, 36, 37
interface, 45
international law, 58
investment(s), 12, 17, 18, 34, 36, 55
Iran, 8, 65
issues, 2, 3, 24, 28

J

Jefferson, Thomas, 53, 68

K

kill, 4, 62
knots, 64

L

laptop, 23
lasers, vii, 1, 2, 3, 4, 5, 6, 7, 8, 10, 12, 14, 15, 16, 18, 20, 21, 25, 28, 29, 30, 31, 40, 41, 45, 48, 56, 57, 58, 59, 60, 63, 67, 69, 70
laws, 58, 63
LCSs, 20
lead, 3, 30, 42, 45, 52, 62
leadership, 17, 23, 55
learning, 19, 48
lens, 72
light, 6, 7, 8, 9, 26, 27, 40, 41, 47, 51, 53, 54, 62, 63, 66, 68
light beam, 9, 41
logistics, 52, 63, 64
low temperatures, 51
LSD, 49
LSI, 42

M

magazines, 18, 22, 32
magnetic field, 51
magnitude, 85
management, 15, 26, 28, 30, 34, 37, 45
manufacturing, 10, 13, 61
Marine Corps, 13, 16, 17, 18, 19, 22, 23, 55
marine environment, 7, 13, 14, 26
marriage, 48
Maryland, 68
mass, 55, 57, 75, 80
materials, 11, 18, 41, 63
matter, 30, 57
media, 68
Mexico, 68
military, vii, 1, 2, 3, 13, 14, 18, 19, 33, 34, 35, 36, 46, 47, 57, 58, 59, 62, 63, 69
mission(s), 15, 16, 17, 19, 22, 56, 59, 64, 65
MLD, 13, 20, 26, 27, 46, 47, 48, 49, 50, 65, 68
models, 9, 52, 78
modifications, 45

N

nation states, 19
National Defense Authorization Act, 32, 34
national security, 30, 36, 37, 59
navigation system, 48
neglect, 74, 76

O

officials, 12, 30, 33, 59, 62
operations, 5, 7, 8, 10, 11, 14, 15, 18, 20, 27, 36, 38, 39, 56, 59, 62, 64, 65
ownership, 18, 61

P

Pacific, 40, 47, 68
permission, 12, 39
permit, 4, 5, 6, 10, 25, 26, 27, 59
Persian Gulf, 1, 2, 9, 12, 20, 21, 66
photons, 51, 68
platform, 18, 42, 48
PMS, 40, 48, 67
polarization, 74
policy, 25, 56, 57, 58, 59, 69
policy makers, 25, 59
pollution, 7
power generation, 30
president, 12, 66, 67
principles, 55, 64, 65
procurement, 3, 26, 41, 42, 61, 62
project, 12, 13, 14, 52, 53
propagation, 71, 72, 73, 78, 80, 81, 82, 84, 85
prototype(s), 1, 2, 10, 11, 12, 13, 15, 20, 21, 22, 36, 41, 43, 52, 55, 56, 65

R

race, 63
radar, 5, 8, 11, 15, 48, 63
radio, 18, 54
recognition, 57
reform, 62
refractive index, 72, 74, 75, 79
refractive index variation, 72, 79
reliability, 21, 26
repair, 67
reporters, 21
requirement, 45, 57
requirements, 3, 16, 17, 18, 19, 33, 42, 51, 55, 59
research funding, 33
researchers, 63
resources, 13, 21, 23, 25
response, 7, 36, 62
risk(s), 6, 8, 14, 45, 55, 59, 63
rules, 19, 59

S

safety, 32, 49, 59
saturation, 7, 62
scaling, 11, 14, 15, 51, 52
scatter, 7
scattering, 7, 9, 62, 63, 72, 73, 78, 79, 80
science, 30, 36, 48
scope, 55
Secretary of Defense, 17, 33, 37, 48, 58, 60, 69
security, 37
semiconductor, 40
Senate, 33, 34, 37
sensing, 65
sensor(s), 6, 8, 11, 15, 18, 19, 30, 62
services, 3, 13, 24, 46, 47, 59
shape, 82
shock, 28
showing, 53, 73
silicon, 30
simulation(s), 71, 72, 73, 80, 85
software, 42, 59
solid state, 1, 2, 10, 22, 23, 41, 48, 49
solution, 7, 26
specific heat, 75
speed of light, 22, 74
spending, 34

state(s), 9, 12, 14, 16, 19, 24, 30, 31, 32, 33, 35, 42, 44, 47, 48, 49, 52, 53, 57, 59, 63, 65, 67, 68, 69, 72
stress, 5
structure, 37
submarines, 3
surveillance, 18
susceptibility, 8, 9
symmetry, 74

T

tactics, 3, 65
target, 4, 5, 6, 7, 8, 9, 13, 14, 15, 18, 41, 45, 47, 48, 49, 57, 58, 62, 63, 65, 69, 71, 72, 73, 78, 79, 80, 81, 82, 83, 85
Task Force, 39, 58, 69
teams, 21
technologies, 2, 16, 17, 18, 19, 24, 30, 31, 33, 34, 36, 37, 49, 55, 60, 64, 65
technology, 10, 14, 17, 18, 22, 23, 24, 26, 30, 31, 33, 34, 35, 36, 37, 41, 48, 52, 55, 60
temperature, 51, 75
testing, 12, 18, 21, 22, 23, 24, 47, 49, 52
thermal energy, 41, 45, 47, 51
threats, vii, 3, 18, 19, 21, 22, 36, 38, 47
time frame, 2, 14, 22
TRA, 69
trade, 21
training, 57, 58, 69
transactions, 37
transmission, 13, 26, 27, 41, 47, 51
treaties, 69
turbulence, 7, 9, 63, 72, 78, 79, 84

U

U.S. policy, 57
uniform, 73, 80, 82
United States, 33, 35, 36, 56, 57, 58, 63, 69
universities, 51
USA, 48, 49

V

vacuum, 63, 74
validation, 13, 14, 56
vapor, 7
vector, 74
vehicles, 12, 13, 15, 18, 60, 61, 62, 67
velocity, 71, 72, 75, 77, 82
vessels, 21, 30, 48
vibration, 9, 28
vision, 2, 14, 16, 17, 57, 58, 62

W

war, 16, 57
war crimes, 57
Washington, 30, 66, 69
waste, 5, 15, 28, 41, 47, 51
waste heat, 5, 15, 28
water, 7, 11, 12, 13, 39, 62
water vapor, 7, 39, 62
wavelengths, 7, 26, 27, 52, 62, 63
weapons, 3, 4, 5, 6, 7, 9, 15, 16, 18, 19, 20, 21, 22, 24, 25, 28, 30, 32, 36, 48, 49, 56, 57, 58, 63, 64, 65, 69
weapons of mass destruction, 65
web, 54
welding, 10, 41, 67
wind speed, 73, 80, 82
Wisconsin, 68
world order, 35
World War I, 35

Y

Yale University, 68
yield, 75
ytterbium, 41